THE

EVERYTHING®

PARENT'S GUIDE TO COMMON CORE MATH: GRADES 6–8

Dear Reader,

As teachers we realize that once parents have students in a middle school setting, they often feel disconnected from the school and what their child is learning. Sure, some schools do a great job of keeping parents informed and involved, but other schools view this time as an opportunity for your child to gain a level of independence and learn the skills of self-advocacy. Regardless, it is important for you to know what standards and skills your child is expected to become proficient with, and how education is changing as a nation.

Math often suffers a negative stigma as being something that people are either good at or not good at. The truth is everybody is capable of some pretty terrific things if they apply themselves and have the necessary supports and tools in place. Only then can the strongest of the naysayers truly see math as the exciting subject that it is.

We hope this book helps you work through the core concepts of middle school math and allows you to gain insight to the changes of the Common Core Standards. If the math your child is learning doesn't come easy, that is okay, as it wasn't designed to. With a little bit of support and practice, everything is possible.

Jamie Sirois and Adam Wiggin

Welcome to the EVERYTHING® Series!

These handy, accessible books give you all you need to tackle a difficult project, gain a new hobby, comprehend a fascinating topic, prepare for an exam, or even brush up on something you learned back in school but have since forgotten.

You can choose to read an Everything® book from cover to cover or just pick out the information you want from our four useful boxes: e-questions, e-facts, e-alerts, and e-ssentials. We give you everything you need to know on the subject, but throw in a lot of fun stuff along the way, too.

We now have more than 400 Everything® books in print, spanning such wide-ranging categories as weddings, pregnancy, cooking, music instruction, foreign language, crafts, pets, New Age, and so much more. When you're done reading them all, you can finally say you know Everything®!

PUBLISHER Karen Cooper
MANAGING EDITOR, EVERYTHING® SERIES Lisa Laing
COPY CHIEF Casey Ebert
ASSISTANT PRODUCTION EDITOR Alex Guarco
ACQUISITIONS EDITOR Brett Palana-Shanahan
SENIOR DEVELOPMENT EDITOR Brett Palana-Shanahan
EVERYTHING® SERIES COVER DESIGNER Erin Alexander

THE EVERYTHING® PARENT'S GUIDE TO Common Core Math: Grades 6–8

Understand the new math standards to help your child learn and succeed

Jamie L. Sirois, MEd, and Adam A. Wiggin, MST

Avon, Massachusetts

An Everything® Series Book.
Everything® and everything.com® are registered trademarks of F+W Media, Inc.

Published by
Adams Media, a division of F+W Media, Inc.
57 Littlefield Street, Avon, MA 02322. U.S.A.
www.adamsmedia.com

ISBN 10: 1-4405-8357-9
ISBN 13: 978-1-4405-8357-5
eISBN 10: 1-4405-8358-7
eISBN 13: 978-1-4405-8358-2

Printed in the United States of America.

10 9 8 7 6 5 4 3 2 1

Library of Congress Cataloging-in-Publication Data

Sirois, Jamie L., author.
Everything parent's guide to common core math: grades 6–8. / Jamie L. Sirois, MEd, and Adam A. Wiggin, MST.
pages cm -- (An everything series book)
Includes index.
ISBN 978-1-4405-8357-5 (pb) -- ISBN 1-4405-8357-9 (pb) -- ISBN 978-1-4405-8358-2 (ebook) -- ISBN 1-4405-8358-7 (ebook)
1. Mathematics--Study and teaching (Middle school)--Standards--United States.
2. Education--Parent participation--United States. I. Wiggin, Adam A., author. II. Title.
QA13.S55 2014
510.71'273--dc23

2014033819

Illustrations by Eric Andrews.

Contents

Acknowledgments

I would like to thank my family, especially my husband Tim, and son Josh, for their unconditional love and support as I ventured into this professional opportunity. Without their sacrifice and constant encouragement I would not have been able to complete this project. Thank you to my fellow sixth grade math teachers, whose knowledge, creativity, and talent encourage me to try new things. I would also like to thank Michelle Morton-Curit, my high-school geometry teacher, for inspiring me to become a math teacher and Dr. Irv Richardson for opening doors to professional possibilities I never knew existed.

—Jamie Sirois

I would like to thank my wife Sarah, as well as my boys Liam and Jacob for their infinite patience as I spent the summer of 2014 staring at a computer screen. I would also like to thank Dr. Beverly Ferrucci for refusing to give up on a young student of education back in Keene. None of this would have been possible without all of your support.

—Adam Wiggin

Introduction: Exactly What Did They Do to Math?!

GOOD FOR YOU. You bought this book. You have proven yourself as being a concerned parent who wants to do everything you can to help your child succeed in middle-school math. Maybe you remember it as a dark time and you want your child to have a better experience. Maybe you were a "rock star" at math and want to make sure that your children follow in your footsteps. Either way you're getting involved, and parent involvement is one of the largest positive factors in a child's success in school.

However, things have changed significantly in the middle-school classroom in the past few years. You may remember having heard the twentieth-century term "new math." And thought to yourself, *"How can math be new?"* It's not. The math has not changed at all. What has changed is the approach on how the material is being delivered to the students from the teachers. "New math" was a nickname given to the paradigm shift when teachers got away from lecture-based, worksheet-riddled classrooms and put the onus on the student. The major focus is having students truly learn the material—to retain it, rather than just regurgitate it on the next test only to be forgotten shortly thereafter.

The unveiling of the Common Core State Standards is the most recent shift in the mathematics classroom (as well as in Language Arts). A heavy focus has been placed on the specific concepts that are taught at each grade level as well as the mathematical practices that should be incorporated into daily learning. Common Core State Standards are looking to improve both content and instruction in classrooms across the United States. For the states that have adopted the program, it provides a common layout of what students should know before proceeding to their next year. What a student is learning in one state is exactly what a student in a different state is learning. Theoretically, if a family were to move across the country in the middle of the school year, there should be a seamless transition for the school-aged children into their new classrooms.

This book will serve as a resource for both the math-loving as well as the math-fearing parent. It will be an aide, to not only become familiarized with the new sequencing of the curriculum, but also to learn how to extend your child's classroom into your home.

There is something to be said for practice through repetition. After all, *"Practice makes progress!"* However, this book will focus more on being proactive about the material. A student who knows what is coming before it arrives in the classroom will have a HUGE advantage over his or her peers. You *will* find an assortment of practice problems throughout this book but let it be known, by no means is this reference intended as a textbook. The sample questions will touch upon some of the most important and tricky concepts of middle-school math but it will not replace the totality of what students experience in middle school.

Thumb through the pages. Take a look at what's on the horizon. Do not try to learn it all at once. Remember, this represents *three years'* worth of mathematics. When you're ready to delve into it a little deeper . . . into the specifics, work alongside your child—not in isolation. They are the reason why you now own this book! Keep an open dialogue with him or her about what they are working on in class, what is going well, what *isn't* going well . . . it is all part of that beneficial classroom extension. We hope that this book serves its purpose as a go-to during those conversations.

PART 1

The Common Core State Standards

CHAPTER 1

What Are the Common Core State Standards?

The Common Core State Standards are a set of national academic standards and instructional guidelines for schools to follow when teaching mathematics. They outline what a student should know and be able to do at the end of each particular grade level, but do not tell teachers how they should teach. The goal of the standards is to create a level playing field of information all students should learn at each grade level, no matter what state, no matter what school. For many years, states have been setting their own expectations of what students should know and be able to do, but when you try to evaluate how the nation ranks in education, the lack of consistent standards makes it almost impossible to compare. This chapter will provide you with an introduction to the Common Core State Standards and identify how this set of standards is intending to improve our educational system.

What Is the Purpose of the Standards?

The purpose of the Common Core standards is to develop a set of common expectations and goals for all students. These standards are intended to prepare individuals for future success in life, in college, or in the workforce. If students are to be ready to compete for jobs, they need a set of clear, consistent, rigorous goals.

The development of the Common Core State Standards began in 2009, when state and national authorities realized the nation's students were not making progress at a rate equivalent to international peers. At this point in time, the state school chiefs and governors that make up the Council of Chief State School Officers (CCSSO) and the National Governors Association Center for Best Practices (NGA Center) began working together to establish the Common Core State Standards. In collaboration with teachers, school chiefs, administrators, and educational experts, the framework for the standards came to life.

The development of the standards was a "backward design." The members started with college and career expectations and worked down through the grades. Working from the top down increased the level of expectation at each grade level. The standards are research based and evidenced based, include application of knowledge through higher-order thinking skills, build upon strengths of current state standards, and include information from top-performing countries so all students can be successful in global job markets.

For more information and a great informative video about the purpose of the standards, please visit the following link and watch the "Three-Minute Video Explaining the Common Core State Standards." This is a nice condensed version to share with others: *www.youtube.com/watch?v=5s0rRk9sER0*.

Setting consistent learning objectives allows the nation to be measured and compared to other countries, as well as to compare states to states. For states that have adopted the standards, it will be easier for students who move from one state to another. Teachers will now know what a child will have been expected to learn in their earlier years of schooling, and it is no longer a guessing game.

Corestandards.org states that "Forty-three states, the District of Columbia, four territories, and the Department of Defense Education Activity (DoDEA) have adopted the Common Core State Standards and are moving forward with their implementation."

The Common Core Standards require teachers to make a change in their professional practice. In order for the United States to compete with international markets, a change is required in what and how math is taught. With the Common Core State Standards, math classes will cover fewer concepts but in greater depth, and a focus on how and why students arrive at an answer will be at the heart of learning. The goal is for students to learn math and be able to apply it to the real world, beyond the walls of the school. The standards identify what students should learn, but does not outline how teachers should teach. Classrooms in schools will still look different from one another because every teacher will take his own approach to how he instructs his students.

What Are the Key Shifts in Mathematics?

Understanding how the standards differ from previous standards, and the necessary shifts they call for, is essential to implementing the standards correctly. When the Common Core State Standards were designed, they were built around three major mathematical shifts: Focus, Coherence, and Rigor. It is important to understand each of these shifts in order to support students in their education.

Focus

The standards call for greater focus in mathematics. What this means is that instead of racing to cover topics in today's mile-wide, inch-deep curriculum, teachers use their power to narrow and deepen the way time and energy is spent in the classroom. The standards require depth instead of breadth, students developing strong mathematical foundations with basic concepts and solid conceptual understanding, which means they understand why and how the math works. The standards also require that students really understand the procedures related to the skills, and can fluently use these skills to solve problems.

When looking at each grade level, a deep focus is on the major work. In grades K–2, teachers focus on concepts, skills, and problem solving related to addition and subtraction. In grades 3–5, the focus changes to concepts, skills, and problem solving related to multiplication and division of whole numbers and fractions. In grade 6, the focus is on ratios and proportional relationships and early algebraic expressions and equations. In grade 7, ratios and proportional relationships, as well as arithmetic of rational numbers, becomes the focus. Finally, in grade 8, linear algebra and linear functions becomes the driving force.

This level of focus on the major areas will help students develop a solid understanding of concepts and the ability to use what they have learned to solve real-life problems they encounter along the way.

Coherence

Coherence is all about the glue that holds everything together. It is about linking information from one grade to the next and building on students' prior knowledge. The standards have been designed around coherent progressions from grade to grade and in order to maintain coherence, it is important that a common language be used at each and every grade level.

Coherence is also maintained through the ways in which the major topics are reinforced. Instead of teaching individual skills and topics in isolation, it is about embedding skills into grade-level word problems that allow for application and reinforcement, so students can make their own connections to the information and become stronger problem solvers. Teachers are being challenged to develop problems that integrate several skills into one complex problem and challenge their students to use critical thinking skills to solve them.

Rigor

Many parents believe that rigor is about making math harder and more difficult to understand, or introducing more advanced topics at earlier grades, but this is not true. Rigor is created through learning tasks that require students to think critically, creatively, and actively about a problem. Rigor is a component of a classroom environment that creates excitement, curiosity, and enthusiasm in students.

Corestandards.org says that by "pursuing conceptual understanding, procedural skills and fluency, and application with equal intensity," learning becomes rigorous. Conceptual understanding is developed by allowing students to access concepts from a number of perspectives, so that math becomes more than memorizing a set of procedures and rules.

Students need to have math facts from 0—12 memorized to have access to more complex concepts and procedures. If students struggle with recalling math facts, the focus for the student is redirected to basic functions, and is taken away from the grade level concepts they should be learning.

Rigorous learning also looks at students having procedural skills and fluency. This is all about knowing what to do, and how to do it quickly and accurately. Students must practice core functions, or the rote skills of math, such as single-digit multiplication, and become experts with the process so that when things become more challenging, the basic skills are not holding students back. Fluency must be a part of every classroom, and can be supported at home through the use of timed fact tests, flash cards, online sites, and practice workbooks. Through the use of these additional resources, students are given more time and more practice to achieve the necessary levels of fluency. Some great websites that support skill development and fluency are:

- *www.xtramath.org*
- *www.wildmath.com*
- *www.freerice.com*
- *www.mathplayground.com*

- *http://aplusmath.com/Games*
- *www.aaamath.com*

There are many sites out there. Feel free to complete a search and see what comes up. If accessing technology is a challenge, don't be afraid to use the traditional flash card activities at home or on a long car ride. It can even become a race to see who can answer the facts more quickly, to add an element of competition and fun!

The last component that creates rigor in a math classroom is the level at which a teacher requires students to apply the math they are learning. The Common Core State Standards expect students to use math in situations. For many teachers it means changing how they prepare for their classroom. Instead of thinking *skill, skill, skill, application*, a teacher should be thinking in reverse: How she can begin with an application problem for students to solve in her classroom, and have her instruction support the students' needs. For example, a problem related to grocery shopping could involve ratios and rates, decimal operations, and data collection. By embedding these questions into the one application problem and asking students what is needed to solve this problem, they might identify the skills and begin to work toward solving the problem. They also may not be able to identify what they need to do, and that is an opportunity to teach the skills of breaking a problem down, identifying key information, and also teaching the skills necessary to solve. Why waste time teaching students things they already know? This, too, helps to create rigor!

Asking students to apply what they know and are able to do to solve real-world problems further solidifies their overall understanding of mathematics. Students become active members of the learning environment, demonstrating their abilities through the mathematical tasks, and meeting the academic demands of the Common Core State Standards.

Many people have different opinions about the Common Core Standards, ranging from full support to challenging every aspect of them. As a parent, it is up to you to become educated about the standards and make an educated, informed opinion for yourself.

CHAPTER 2

What Are the Eight Standards for Mathematical Practices?

Math is more than a set of concrete procedures and rules. It is about developing habits of mind that will follow a student and allow her to be successful in her future. The Common Core State Standards have outlined eight mathematical practices that are to be integrated into daily lessons. These are a set of skills that a math teacher hopes to develop in his students so they can find long-term success in life and see how math is integrated in the world around them. This chapter will detail each of the eight standards and explain how you can help support the development of these standards at home.

Standard 1: Make Sense of Problems and Persevere in Solving Them

Teachers are finding that students want to give up when something becomes too challenging. This standard takes the opportunity away. Students are now expected to look at a problem, break it down into parts that they understand, and begin working with the most basic piece. This is referred to as the problem's *entry point*.

It can be hard to motivate your child to want to work through the problem and take the time to wait for her to figure out how to approach it. So often parents and teachers want to swoop in and help, but the second you pick up the pencil and show her how to do it, the learning opportunity for the child is lost. Allowing her enough time to work though tough problems develops a process for problem solving.

One approach to problem solving is to use a four-step plan. This consists of the following steps:

1. Understand
2. Plan
3. Solve
4. Check Your Work

Rather than just diving into the problem blindly, students are encouraged to take a step back and develop an approach to solving the problem, beginning with the most basic piece.

What Does It Mean to Understand a Problem?

Understanding a problem is being able to break the problem down and determine the following pieces:

- What facts do you know?
- What do you need to find out?

Strategies to help answer these questions are:

- Read and underline key points.
- Cross out useless information so it doesn't confuse you.
- Box in or circle key information that you want to use.
- Look up vocabulary words you do not know the definition of.

Once you help your child develop this process, understanding what a problem is asking you to do becomes much easier.

How to Plan to Solve a Problem?

Developing a plan to solve a problem is all about deciding what strategy to use to answer the question. There are many strategies available, and the information provided in the problem will help to guide your child as to which approach is best. Here are some strategies that can help:

- Draw a picture
- Make a chart or diagram
- Guess and check
- Make a list
- Make a simpler version of the problem
- Work backward
- Look for patterns
- Create a graph
- Use an equation

Now It's Time to Solve

Solving a problem is all about following through with whatever strategy you've picked. This is where a student shows all of his work, whether it is writing and solving an equation, drawing a picture, or any of the other number of strategies mentioned. The best way to support your child is to help him organize his work. Help him to arrange his work neatly, whether horizontally or vertically. You could also make boxes for him to show different parts of the problem so that all parts of that particular step are together. Finally, remind him to label the important numbers using appropriate units, and circle the final answer.

Check Your Work

Ahhhh . . . now you can give a big sigh of relief. The problem is solved; however, you are not done yet. Successful, strong math students always take the next step of asking themselves, "Does the answer make sense?" Asking this simple question makes a world of difference between successful math students and those who may struggle. By taking the extra minute to answer this question, students can review their work and look for mistakes. Encourage your child to check her work by doing the inverse or "opposite" operation. For instance, if she did multiplication to find the answer, have her take the product and divide it by one of the numbers used in the multiplication problem. If the answer is the other factor, then her multiplication is correct. If the answers differ, she should go back and repeat the solving process.

Helping students to solve problems using various strategies, check their work, and determine if the answer is reasonable will lead to long-term success. As a parent, be there to help with organization, make reference to various strategies, and encourage them to check their work, but do not take over. Always lead with questions, but allow students to make the progress on their own. Try not to become impatient with your struggling student. Encourage and support the struggle, and know that this struggle leads to future success. Good problem solvers try, fail, regroup, and try again.

Standard 2: Reason Abstractly and Quantitatively

How many of you are reading this title and thinking, "Oh my, these words are already confusing"? This mathematical practice asks students to contextualize numbers, decontextualize words, and use reasoning to solve problems. If this still doesn't make sense to you, you are not alone. Let's make it a bit easier to understand.

To reason abstractly means to be able to decontextualize a problem or to be able to pull numbers from a word problem out of context, show it symbolically using pictures and/or diagrams, and turn it into a set of numbers and symbols that can be worked with mathematically.

How do I decontextualize a problem?
Consider this word problem: Mary ran 3.5 miles per hour for 2 hours. How far did she run? Decontextualizing this problem turns it into $3.5 \times 2 = 7$. However, there is no meaning to the answer, so let's look at it with labels: 3.5 mph × 2 hours = 7 miles. Mary is able to run 7 miles in 2 hours' time at a rate of 3.5 mph. We have taken a word problem, interpreted the meaning of it, assigned numbers and operations, and carried out the necessary steps to solve. We have reasoned abstractly.

Abstract reasoning also means to contextualize a problem, which is to create meaning given a set of numbers and symbols, and to put them into a real-world context. For example, if given the math problem $3.99 \times 5 = 19.95$, a real-world situation could be, Johnny bought 5 gallons of milk at $3.99 each; how much did Johnny spend? The answer would be that Johnny spent a total of $19.95.

Quantitative reasoning is all about applying the math that a student is learning to everyday life. This is really an essential skill set to learn. This skill asks students to look more deeply into mathematics and view problems as not just a set of calculations, but as an opportunity to analyze a mathematical situation and understand the meaning of the quantities or numbers being used. It requires students to be able to interpret graphs, models, and tables, and draw conclusions from their observations. This skill set includes estimating, checking if answers make sense, looking for other ways to solve a problem, and, most importantly, knowing the mathematical concepts well enough to use them in varied situations.

How can I help my child apply this skill set at home?
The best way to support the development of reasoning skills is to ask questions of your child. Questions could include the following: Why did you choose to solve the problem this way? What does the answer mean? Can you explain to me what this graph is representing? Ask your child to teach you about the math they are learning, and explain the details he knows and understands.

Being able to reason abstractly and quantitatively allows for students to access just about anything, because they know how to make connections between ideas that are not obvious, they can transfer knowledge from one application to another, and can draw conclusions based on given information. These are skills that are used in everyday life.

Standard 3: Construct Viable Arguments and Critique the Reasoning of Others

How many of you have children that like to debate a statement that you have made? Many of you are probably nodding your head in agreement. If this is the case, the third standard for mathematical practice is right up your child's alley.

Here students get to make their own conjectures, or statements that are based on their opinion of provided evidence from a problem. They act as a lawyer representing a case, analyzing the details and determining if the information is justified. Is there a counterexample that proves the theory wrong? If so, what is it? They need to look at the problem from both sides, acting as both a prosecutor and a defense attorney. Is the answer or explanation provided accurate and logical, or is there a mistake? If there is a mistake, students need to be able to explain what it is and how to fix it.

This standard also teaches students how to respectfully communicate ideas and opinions, whether in agreement or not. It sets up an environment where ideas can be challenged, but students are not threatened. Most importantly, students must justify and explain why their conclusion (answer, explanation, statement, etc.) is correct. They must also learn to listen to others, process the ideas of others, and determine how these statements compare and contrast to their own opinions. Students are encouraged to ask questions of one another and to look for more than one way to arrive at an answer. When someone challenges the thoughts and ideas of another, this leads to rich mathematical discussion, and further develops students who are mathematically proficient.

Standard 4: Model with Mathematics

Many people interpret this standard as strictly using manipulatives (tangible objects) to build a physical representation of the math being presented; however, it is so much more. Modeling with mathematics includes everything from building a physical representation using tangible objects to writing an equation to represent a situation. It includes analyzing information and presenting it in tables, graphs, flowcharts, diagrams, and even using graphing calculators to make a model.

The purpose of this standard is to allow students to gain a level of comfort with various modeling tools, and to be able to move flexibly between these tools depending on what question is being asked.

For example, a basic sixth-grade problem may ask students to determine which is the better buy, 10 raffle tickets for $1.00 or 2 raffle tickets for $0.25? A visual picture may be drawn to help students make meaning of the math. Refer to the following diagram. Do you see how this might be helpful to a student?

Option 1: Purchase 10 tickets for $1.00

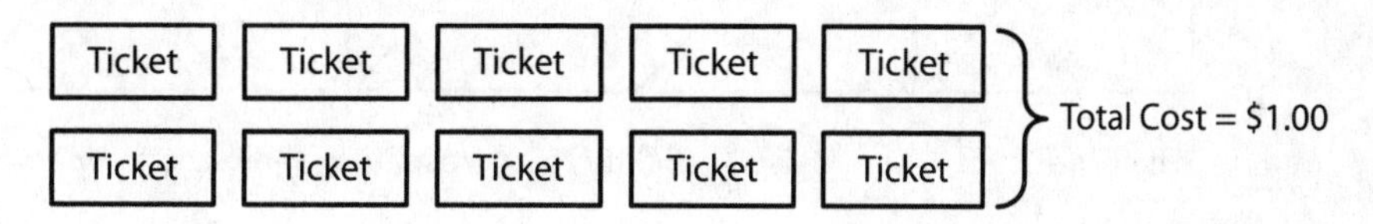

Option 2: Purchase 2 tickets for $0.25

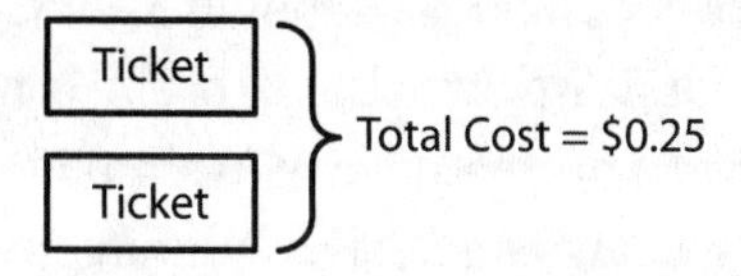

Five groups of two tickets makes 10 tickets . . . so five groups of $0.25 is the total cost.

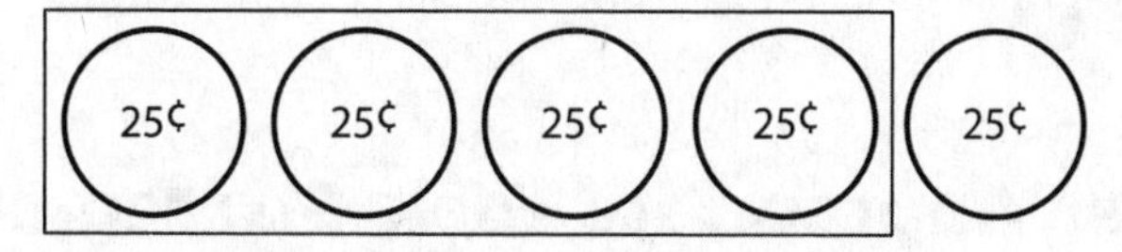

$1.00 + $0.25 = $1.25

Using a model or picture to represent what is happening in the problem allows for students to make meaning and draw a conclusion. In this particular example, it is cheaper to buy 10 tickets for $1.00 because if they are purchased two at a time, the total cost for 10 tickets is actually $0.25 more.

Other ways to model mathematics include building an area model to show multiplication of two fractions. Building area models helps to develop conceptual understanding. Look at the area model created to show multiplication of two simple fractions.

$\frac{3}{5}$ is represented by 3 out of 5 columns being marked using the backward diagonal.

$\frac{1}{2}$ is represented by 1 out of 2 rows being marked using the forward diagonal.

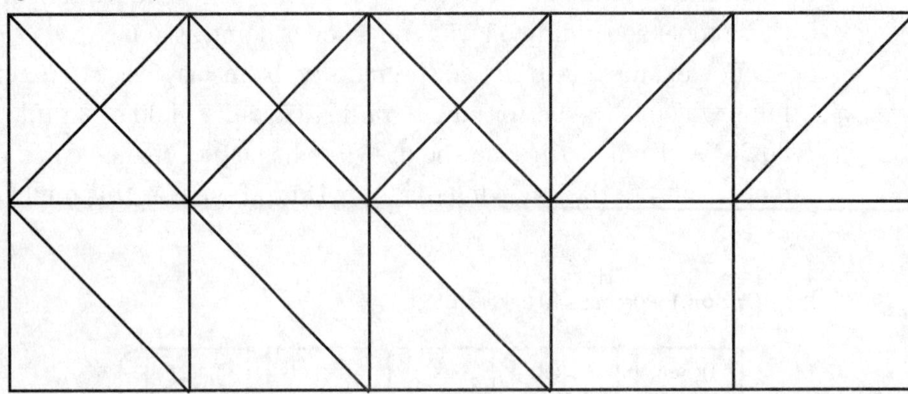

The product of $\frac{1}{2}$ and $\frac{3}{5}$ is $\frac{3}{10}$ and this is represented where the forward and backward diagonals overlap, or where the X appears in each rectangle. If you count, 3 out of 10 rectangles contain an X, so $\frac{3}{10}$ is the product.

These are two specific examples of modeling, but are not the only forms that exist. Modeling with mathematics involves many of the other standards for mathematical practices. It may require students to simplify complicated situations, analyze information to draw conclusions and make conjectures, interpret their math results, and provide a context for the situation. It even expects them to answer the question, "Does my answer make sense?"

Standard 5: Use Appropriate Tools Strategically

A mathematically capable student will approach a problem and consider all the possible tools that are available to him. Mathematical tools take on many forms. These tools can be as simple as pencil and paper or as complex as the most innovative technology. The purpose of a mathematical tool is to help a student solve problems and make sense of what is happening. The focus of this standard is not which tool they choose to use, but how they use it.

Students need to understand that every tool has a purpose, a benefit, and oftentimes a limitation. Here is a list of tools to consider when supporting your child:

- Paper and pencil
- Physical objects such as cubes, geometric shapes, place value manipulatives, fraction bars, etc.
- A ruler and/or protractor
- A calculator
- Spreadsheet software such as Microsoft Excel or Google Sheets
- Computer software for algebra, statistics, or geometry
- Drawings or diagrams (number lines, tally marks, tape diagrams, arrays, tables, graphs)
- Scissors, tracing paper, graph paper
- Web-based resources

Having some of the more basic tools on hand at home is helpful so that if a student thinks of a way to solve a problem requiring tracing paper or graph paper, for instance, he can easily access this tool and continue with his work.

Again, as a parent you may be interested in jumping in to save the day with a suggestion of a tool, but your job is very similar to that of a teacher. Provide an opportunity for your child to pick the tool he feels is appropriate, and then have the discussion about the benefits and limitations of that choice. This will allow his understanding to grow and develop, and will help him become a stronger math student.

It is easy for a child to always want to use a calculator, and although this can be a useful tool, using it to do basic computation actually puts him at a disadvantage. Encourage your child to do mental math for basic addition, subtraction, multiplication, and division computations, and save the calculator for more complex material.

As your child becomes more comfortable using a variety of mathematical tools, you can refine her tool-selection method by asking some simple questions. First, is the tool necessary? Is the tool easy to use? Does the tool make the problem-solving process more effective and efficient? Lastly, does the tool support the mathematics in a meaningful way? Modeling this process with her will lead to her being able to ask these questions of herself. Little by little, her problem-solving ability will become stronger and you will see the amount of support required at home diminish.

Standard 6: Attend to Precision

Mathematically capable students look to communicate accurately and effectively with others. Accuracy is more than just arriving at the correct answer. Accuracy includes communicating mathematics effectively by using appropriate definitions that support the problem. Attending to precision means choosing the right symbol or set of symbols when writing an expression or equation.

Surprisingly, many students use the equal sign incorrectly. For example, if students were asked to add a string of numbers together, like $2+3+7+9+11$, they might take two numbers at a time, like this: $2+3=5$, and then add on the next number so that the math ends up looking like $2+3=5+7=12+9=21+11=32$. This work is mathematically incorrect. By having a set of equal signs between each of the expressions, you are saying they are all equal to one another. This is not true. $2+3$ does not equal $5+7$ because 5 is not equal to 12; nor does $5+7=12+9$ because 12 is not equal to 21. The correct way to write this mathematically is $2+3=5$, $5+7=12$, $12+9=21$, and $21+11=32$, so the sum of $2+3+7+9+11=32$. Notice how commas are separating the individual math calculations; this is mathematically correct.

Other elements that are included in attending to precision are paying attention to the units of measure, labeling answers, and labeling axes when graphing. Really, this standard asks students to pay attention to all the little details that go along with learning math. It is the little details that are most difficult for students, because it requires them to slow down, really look at the problem, and think about all the intricate details that lead to a full, well-developed solution.

Attending to precision also focuses on the degree of precision that is needed for numerical answers. Students need to be able to think about when it is appropriate to round, and to what place value should they round to.

If students were asked to solve the following problem, what degree of accuracy should they use?

Mary buys a box of saltwater taffy that holds four pounds for $13.99. How much is she paying per pound?

Since this problem is dealing with money, the degree of accuracy should be the hundredths place, because money is expressed to the nearest cent (two decimal places). Let's look at the math.

$$\$13.99 \div 4 = 3.4975$$

Rounding this to the nearest cent gives an answer of $3.50, so Mary pays $3.50 per pound of saltwater taffy.

Encouraging your child to attend to precision will help him to look for the finer details of all things in life. He will become more observant in his own work and the work of others, and will learn to communicate mathematics efficiently and effectively.

Standard 7: Look For and Make Use of Structure

When you read this, are you thinking that making use of structure in math is simply organizing your work neatly, or that it is building or modeling something? That is not the structure that this standard is striving to develop.

Mathematically proficient students look to find patterns in math. They then use the patterns and extend them to help solve more challenging, complex problems. There is structure in all strands of mathematics. For example, in exploring geometry and the study of quadrilaterals, there is structure in the way each quadrilateral relates to another. Look at the following Venn diagram.

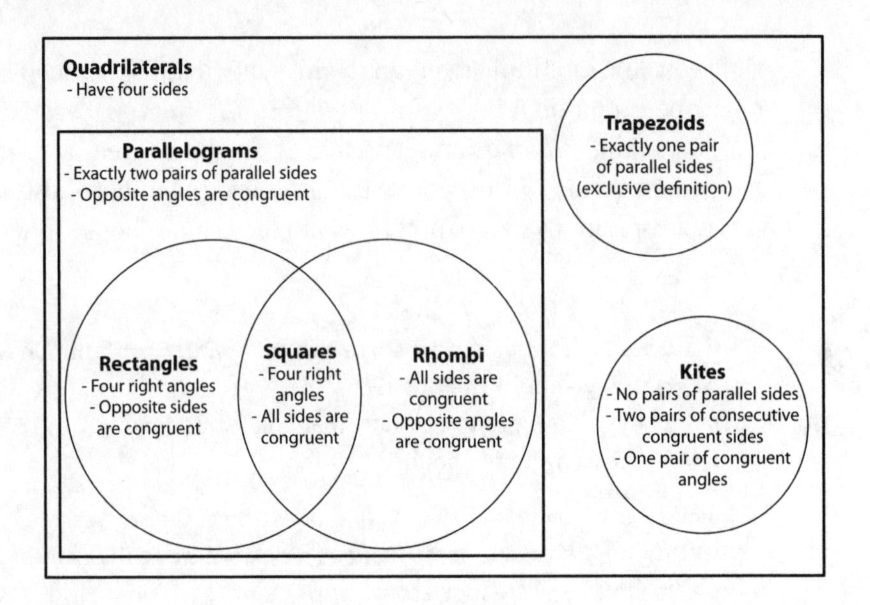

This Venn diagram shows that all rectangles are parallelograms, because the circle of rectangles is fully inside the parallelogram structure. It also shows that kites have nothing in common with trapezoids other than being a four-sided shape because their circles are not connected at all, but both shapes are included in the quadrilateral structure. By using this diagram, students can look for patterns, similarities, and differences among four-sided shapes.

Some textbooks define a trapezoid as having at least one pair of parallel sides. This definition is more suitable for advanced math topics seen beyond middle school. The advanced definition is considered the inclusive definition, because it includes parallelograms in the definition of trapezoids.

A student makes use of structure when stating it is impossible to get an odd-numbered sum when all the addends are even. Here, the student demonstrates an understanding that the sum of two even numbers is always even. A student may make use of structure when working with linear functions, because she sees that the rate of change is constant. Looking for structure in mathematical situations and connecting these observations between topics promotes problem solving and understanding of new mathematical concepts.

Making use of structure also includes looking at a problem from a different perspective. Maybe your child begins to see complex things as a collection of single items or more simple objects. An example of this would be finding the area of an irregular figure, like the one pictured here.

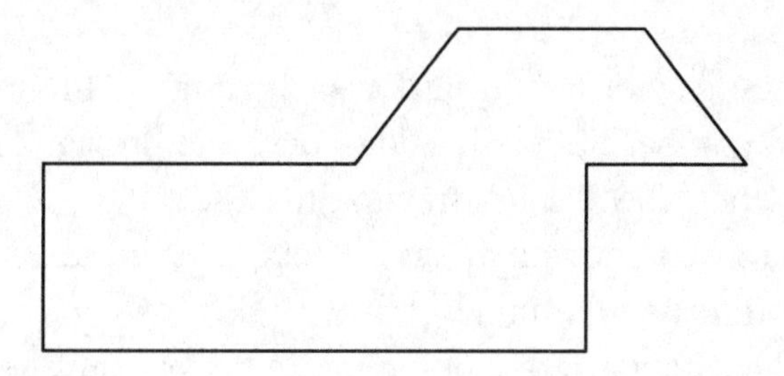

This irregular figure can be broken into a rectangle and a trapezoid, and what seemed like a challenging problem has been made simpler just by looking at it with a different perspective. A student would now be able to find the area of the rectangle and the area of the trapezoid, and add the two together to get the total area of the irregular figure.

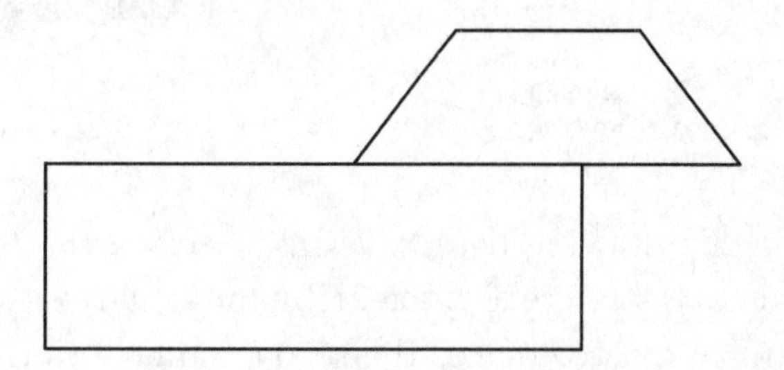

Standard 8: Look For and Express Regularity in Repeated Reasoning

This standard also focuses on recognizing patterns and structure. The goal of this standard is to find a pattern, be able to apply it to solve a single problem, and then use the information gained to make generalizations about how this can be applied to future situations.

For example, the triangular number sequence (1, 3, 6, 10, 15, 21, 28, 36, 45 . . .) is generated from a pattern of dots that form a triangle (see the following picture). This sequence has a pattern. Can you see the pattern?

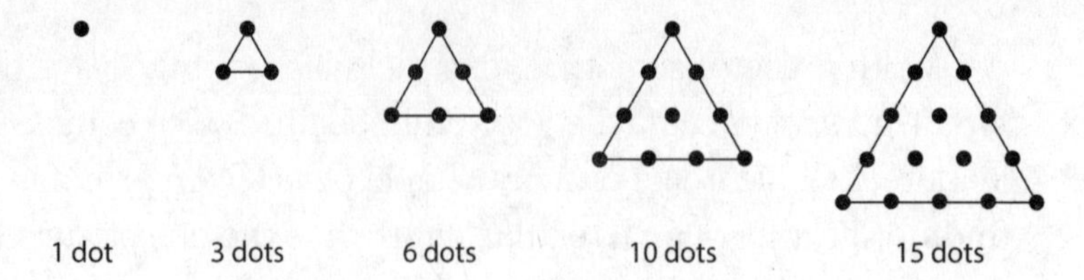

You may observe that for each additional number in the sequence, the same number of dots is added onto the previous image. The first drawing has 1 dot, the second drawing has 3 dots $(1+2)$, the third drawing has 6 dots $(3+3)$, the fourth has 10 dots, and the pattern continues. What is the twelfth triangular number?

The standard of looking for and expressing regularity in repeated reasoning would encourage students to look more deeply into the pattern. What if we reconfigured the drawing and replaced the dots with cubes? Does this make it easier to see a way in which we could generalize what is happening?

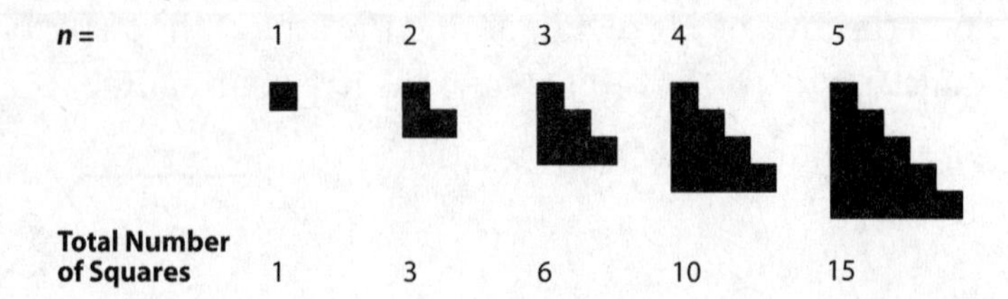

In the drawing here, n is representing the term in the sequence, so when $n=1$, it references the first number in the sequence, or the number 1. When $n=2$, it references the second number in the sequence, or 3, and so on. What would happen if each of these figures were doubled; what would the figure become?

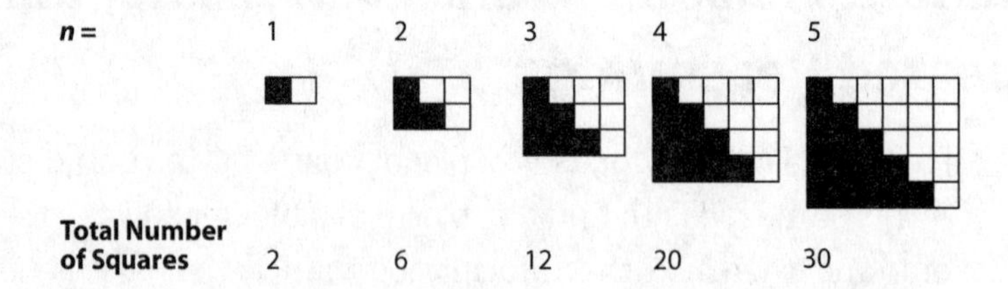

If you look at the image, each triangular number has been transformed into a rectangle. How many squares are there in each figure now? The first number in the sequence now has 2, the second number in the sequence has

6, and the third number in the sequence has 12. Each of these is double the amount of dots in the triangular number sequence.

Looking for an Easier Way

An easier way to find the number of squares in each rectangle is to find the area of the rectangle, or multiply the base times the height. A table is a great way to help organize and generalize information.

Term in Sequence (n)	Base (in units)	Height (in units)	Area (in sq. units)
1	2	1	2
2	3	2	6
3	4	3	12
4	5	4	20
5	6	5	30
6			

What observations can you make? The base is always 1 more than the term of the sequence and could be represented algebraically by $(n+1)$. The height is equal to the term in the sequence, or n. The area of the rectangle is found by multiplying base × height or $(n+1)n$.

Using this, what is the area of the sixth term in the rectangular sequence?

$n=6$ $\quad$ $Base=(6+1)$ $\quad$ $Height=6$

$Area=base \times height$

$Area=(6+1)6$

$= 7 \text{ units} \times 6 \text{ units}$

$= 42 \text{ sq. units}$

How can this information be used to determine the number of dots in the triangular sequence? Each term in the triangular sequence had half of the number of dots as the rectangles, so we need to divide the area in half.

Pulling It All Together

The formula for determining the number of dots in any term of the triangular sequence can be found using the formula: $[n(n+1)]/2$. Does this work?

Earlier you were asked how many dots would be in the twelfth triangular sequence. In this instance $n = 12$, so by substituting in the value for n, the solution will be found.

$[12(12+1]/2$
$[12(13)]/2$
$[156]/2$
78

78 is the twelfth triangular number. This could be verified by drawing each term in the triangular number sequence until you arrive at the twelfth number in the sequence; however, this is not an efficient way to do mathematics.

Using this as a model to help your child look for patterns can help her develop her own set of strategies when faced with a similar situation. It is a process, one that requires being creative and curious. You have to ask yourself the "what if" questions and explore what happens when you try. It may lead you to a correct answer or it may not, but by taking the risks you are doing mathematics.

Connecting the Standards for Mathematical Practice to the Standards for Mathematical Content

The Standards for Mathematical Practice directly relate to the Standards for Mathematical Content. The purpose of both sets of standards is to develop students whom are strong problem solvers and have the ability to connect mathematical ideas. The goal is to engage students with mathematical content through exploration and application.

This collection of standards aims to help students understand why math works the way that it does, and be able to explain how it can be applied in other contexts. Ideally, these two sets of standards are intertwined in the learning process so that students have a strong foundational understanding of a topic and are able to engage in more complex mathematical content.

CHAPTER 3

How Can I Help My Child Be Successful?

Helping your child be successful in mathematics is all about allowing your child to *do* mathematics. Swooping in to save the day by doing the work for your child is not the answer. It will only help to develop a child that is dependent on you for everything, including taking his test for him. The best thing you can do is support your child by following the helpful hints outlined in this chapter.

Note Taking and Homework

In many middle schools, students begin to learn to take notes. These notes consist of definitions, example problems, step-by-step procedures, suggestions for organizing work, and helpful pieces of information to reference. Sometimes notes don't look like the traditional notes you may be familiar with. The Common Core expects students to develop conceptual understanding through models and exploration. Your child's notes may be a collection of diagrams, models, and written responses. Students may be given specific questions to ponder and explore to help develop their understanding, or may have papers from a particular activity they completed in class. Regardless of what they look like, the purpose of the notes and classroom activities are to support student learning.

Encouraging Neat Note Taking

Being able to use the activities from class at home can become a challenge when you or your child cannot read the notes. This is the first step in supporting your child. Help him take well-organized, thorough notes. This can be done using already existing notes. You can sit together at the dining room table and copy the notes over to make them neater. You can reinforce major topics by showing him how to use a highlighter to make key information stand out. You can even use the strategy of making him recopy his notes to help him study information and become better with the note-taking process. Although a grueling task, it can help!

If your child is not up for any of the aforementioned ideas, you could watch a news program, movie, or favorite television show together and each write your own set of notes, then compare the results. Discuss the two sets and share why each of you made the choices you did. What makes a set of notes neat? What helps make information accessible? What information should you write down?

The purpose of these activities is to reinforce the importance of taking "usable" notes. They need to be neat and provide necessary information. Every set of notes should include some basic information:

- Student Name
- Date

- Period
- Heading or title of notes
- Definitions of vocabulary words
- Sample problems

Many teachers have a specific template that they would like their students to use, while others ask the students to find a method that works for them.

Reviewing Notes

How can you support your child in using his notes? This is where you wear the hat of a tour guide.

As your child is completing his homework and becomes stuck on a problem, instead of completing the work for him and saying, "Here, this is how you do it, just copy this down," put on the tour guide hat and guide him through the process of using his notes, textbooks, and resources to determine an answer.

Many tour guides lead with a question: "How many of you have been here before?" Your question might be, "Have you seen something like this before?" Hopefully your child responds with a "yes" and you ask your next question. "Where have you seen this?" Again, your child might respond with, "We did something like this in class." Here is the perfect opportunity to say, "Well, why don't you take out your notes from class so we can see a sample problem, and go from there." Instead of giving him the answer, you directed him to where he can go to get an answer for himself.

A great next step would be to have him take on the role of teacher. Ask him to explain what he learned today and how he completed the sample problems on the page. This strategy further develops his understanding of the topic, because he has to reconnect with the information, make sense of it in his mind, and then explain it in words that you will understand. As he is explaining it, he might get an idea as to how to solve the current homework problem. Even if he doesn't gain information for his homework, he is reviewing key information that he will need to move forward.

Don't get confused by thinking that your child understands something just because a teacher has modeled a problem. There is a difference between being able to do it independently and being able to do it with a step-by-step model.

If, at the end of reviewing his notes, your child still does not know how to solve the problem, take him to the next stop on your tour, the textbook. Encourage your child to use the index to look up the topic he is learning, and read the chapter. A textbook is a great resource because it has additional example problems, helpful hints, and details outlining the steps to go about solving.

This is another key moment when you have to remember that your child must do the work, not you. As difficult as it may be, fight the temptation to step in and solve. You want your child to be an independent, self-motivated learner. You won't be able to help your child when he is struggling in college, so now is the time to let him struggle a little and learn the strategies he needs to be successful.

Teachers want to see what students are able to do independently. If a student comes in with all problems solved and no questions, a teacher is lead to believe that he understands everything. Encourage your child to write the teacher a specific question about the problem or problems he doesn't understand. Include steps that he did try, and write about the steps he took to try and figure it out. Not only does this show he didn't understand it, it shows an incredible amount of effort on his part to try and understand the concept! Teachers love this even more than the right answer.

Using Notes and Homework to Study for a Test

Teachers often tell students to use their notes to review for a test, but what does that look like? To someone who has been through twelve plus years of schooling it may be obvious, but to others, studying notes may be a foreign concept.

Learning Vocabulary

Students can use their notes to study vocabulary words and definitions by making note cards from the information provided. Flash cards serve as a great tool for things that need to be memorized, such as definitions, multiplication tables, symbols, properties, formulas, names of place values, and any other specific details. If students do not know the meaning of these basic things, access to more complex concepts will be limited.

Having a set of different-sized note cards on hand at home makes it easier for a student to use this study tool. If note cards always need to be purchased, chances are students will not get into a habit of making them.

Sample Problems

Students can also use their notes to get sample problems to practice. Teachers model the types of problems they expect students to know and be able to do. Many students will simply want to read over the sample problem and say, "I studied it. I'm all set." But this is not studying. When a teacher says use your notes to study, the expectation is that students will copy problems over that have been completed in class, and try them over again on a separate piece of paper. If the correct answer is reached, fantastic; if not, it is time to look at both sets of work and find where the mistake was made. That is studying.

Homework is another place to access example problems to review and try again. The homework assignments were completed while learning was still happening, so going back and trying them again at a later date, especially before a test, is a great way to measure a child's level of mastery. If he is able to get each and every question he tries correct, chances are he is ready to "rock and roll" on the assessment. If he still makes frequent mistakes, then the next step would be an extra help session with the teacher.

Encourage your child to begin studying days before the test, and not hours. If she just studies the night before and stumbles across questions, it is too late to be able to ask the teacher for help.

Lastly, the textbook is another fantastic place to get extra practice problems to use while studying. Most likely, your child's teacher did not assign all of the problems during the course of the unit. As a parent, you can look through the chapters and pick some problems for your child to try. The great news is, if you pick the odd problems, many textbooks put these answers in the back of the book. Yippee, free answers!

Create Practice Problems

If the notes, homework assignments, and textbooks do not provide enough practice for your child, you can always create problems of your own. In fact, this is a great activity for your child to do. If she is able to come up with a mathematical scenario and solve it correctly, she is setting herself up for success.

Let's say you are hosting a study session at your home some evening. A great suggestion would be to have each person create a mock test for a friend to take. In creating a mock test, students are learning to anticipate the types of questions a teacher might ask on a classroom assessment.

To add a little excitement to the study session, you can even make a game out of learning. *Jeopardy!* is an easy game to create and is a great way to review and prepare. Children don't like to get negative points, so they work really hard to answer each question correctly. You can also make as many or as few categories as you want or need. Make it a family game night and challenge everyone to turn their math brains on.

Model How to Study

How many of your children had to study for math in elementary school? Probably none. If your child went to school, did the class work, listened to the teacher, and completed the homework, he was probably ready for the test. Many of the tests were probably fill in the blank, write your answer in the box type of tests. Not anymore!

Students often mistakenly study the material that they already know and are comfortable with rather than the concepts where they have gaps and are less proficient. This leads to inefficient study sessions, ultimately resulting in frustration after a poor test score.

Since your child has moved on to the middle-school years, a lot is going to change. Your sweet, fun-loving, eager-to-learn child is approaching adolescence . . . hold on tight. But it is true; lots of things are going to change in middle school, including the way in which your child needs to prepare for tests. No longer is just being present in class enough. Your child needs to be engaged in the learning process and engaged in the study process.

How do children learn to study if they have never had to do it before? Teachers cannot come home with your children and show them what to do. Teachers can give ideas, models, tips, and tricks, but unless someone is willing to help with the process at home, it can go in one ear and out the other. This is where you come in.

Sometimes it is embarrassing for a child to teach his parents. If this is the case, try and arrange a time when your child can teach it to someone else in an extra help or study-buddy session.

Your job is to be the model. Do you have something in your own life you need to remain informed about? Maybe you are a nurse and you need recertification hours, or a horticulturist who needs to brush up on some new species of flowers. Maybe you have a new human resource manual that just came out at your place of employment, and you need to know the rules and regulations. Maybe you are going to get your boating license and need to prepare for the licensing exam, so you have to brush up on the boating safety laws in your state. Regardless of your reason for studying, anything you can do to model how to study will greatly help your child. Sometimes just knowing that a parent is sitting in the same room studying helps a child take the necessary steps to prepare for her own test. If you don't have any subject matter to study for yourself, take on the role of a student and ask your child to teach it to you.

Access Tools Online

We live in a time where information is at our fingertips. The digital world has opened up endless possibilities to access information about anything we might possibly want to know. Our job is to figure out how to use the tools to our advantage.

The following list is a collection of websites that are helpful in supporting students. They include both free, unlimited-use sites, and sites that require membership for unlimited use. Please take some time to share these sites with your child, so she can expand her own toolbox and know where to go when trouble arises.

- *www.aaamath.com*
- *www.amathsdictionaryforkids.com*
- *www.aplusmath.com*
- *www.coolmath.com*
- *www.figurethis.org*
- *http://freerice.com*
- *www.funbrain.com*
- *http://illuminations.nctm.org*
- *www.ixl.com*
- *www.khanacademy.org*
- *www.kidsites.com/sites-edu/math.htm*
- *http://mathforum.org/dr.math*
- *http://math.furman.edu/~mwoodard/mquot.html*
- *www.mathhelp.com*
- *www.math.rice.edu/~lanius/Lessons*
- *www.mathplayground.com*
- *http://nces.ed.gov/nceskids/*
- *www.summerskills.com*

- *www.superkids.com*
- *www.visualfractions.com*
- *www.webmath.com*

Supporting your child can look very different from one child to the next. Every child learns differently, needs different levels of support, and thinks about and approaches math in his own way. Some children seem to have a natural talent for mathematics and others struggle every step along the way. Wherever your child lies on this continuum is okay. Do your best to provide what you can. If you really find your child is struggling and you do not feel comfortable that you can teach the mathematics correctly, consider hiring a tutor. Many schools have a list of available tutors through their guidance department, or you can contact a local tutoring agency. Tutors can often provide the added instruction and reinforcement your child needs to navigate his way through math.

PART 2

6th Grade Common Core Standards

CHAPTER 4

What Is Expected of My 6th Grader?

You may remember what you learned as a sixth grader, but times have changed. This chapter outlines the major focus areas that students will spend the majority of their time learning during the school year. Some of it may look a bit different, but don't worry, you will get through it. Much of the content includes pre-algebra concepts and the use of variables to represent unknowns. This may be unfamiliar to your child in the beginning, but by the end of sixth grade she should feel very comfortable with many of these concepts.

What Are the Critical Areas?

There are four areas that sixth-grade students should be spending the majority of their school year learning. These are defined by Corestandards.org as:

1. Connecting ratio and rate to whole number multiplication and division, and using concepts of ratio and rate to solve problems.
2. Completing understanding of division of fractions, and extending the notion of number to the system of rational numbers, which includes negative numbers.
3. Writing, interpreting, and using expressions and equations.
4. Developing understanding of statistical thinking.

Although these are four relatively short statements, they are jam-packed with a lot of mathematics. Let's explore each of the critical areas in more detail.

Critical Area One

Students are expected to take what they know and understand about multiplication and division and apply it to solve ratio and rate problems. The real focus is being able to find equivalent ratios (also can be thought of as equivalent fractions) that are made larger by multiplying by a scale factor, or smaller by dividing by a scale factor.

$\frac{3}{5} \times \frac{5}{5} = \frac{15}{25}$

$\frac{5}{5}$ is the scale factor being used

$\frac{3}{5} = \frac{15}{25}$

$\frac{3}{6} \div \frac{3}{3} = \frac{1}{2}$

$\frac{3}{3}$ is the scale factor being used to simplify $\frac{3}{6}$

$\frac{3}{6} = \frac{1}{2}$

Students also learn to use a multiplication table to identify equivalent ratios and rates. Pairs of rows (or columns) highlight a set of equivalent ratios.

x	1	2	3	4	5	6	7	8	9
1	1	2	3	4	5	6	7	8	9
2	2	4	6	8	10	12	14	16	18
3	3	6	9	12	15	18	21	24	27
4	4	8	12	16	20	24	28	32	36
5	5	10	15	20	25	30	35	40	45
6	6	12	18	24	30	36	42	48	54
7	7	14	21	28	35	42	49	56	63
8	8	16	24	32	40	48	56	64	72
9	9	18	27	36	45	54	63	72	81
10	10	20	30	40	50	60	70	80	90

Look at the highlighted rows. All of the multiples of 4 and all of the multiples of 6 are highlighted. If these are each made into a fraction you would be looking at a series of equivalent fractions.

$$\frac{4}{6}=\frac{8}{12}=\frac{12}{18}=\frac{16}{24}=\frac{20}{30}=\frac{24}{36}=\frac{28}{42}=\frac{32}{48}=\frac{36}{54}$$

This happens with any two rows (or columns) you choose in the multiplication grid.

The multiplication table is a great tool to help students who have difficulty identifying common factors of a pair of numbers. This is a necessary skill to simplify fractions. If students need help, use the table to simplify by moving to the left to find a smaller equivalent fraction. Check to see if this new fraction can be simplified by looking in the rows above. You want to move as far left and up the table as possible.

$\frac{20}{30} = \frac{4}{6}$. $\frac{4}{6}$ can be found in another pair of rows by moving up: $\frac{4}{6} = \frac{2}{3}$. (See following table). This fraction is now simplified.

x	1	2	3	4	5	6	7	8	9
1	1	2	3	4	5	6	7	8	9
2	2	4	6	8	10	12	14	16	18
3	3	6	9	12	15	18	21	24	27
4	4	8	12	16	20	24	28	32	36
5	5	10	15	20	25	30	35	40	45
6	6	12	18	24	30	36	42	48	54
7	7	14	21	28	35	42	49	56	63
8	8	16	24	32	40	48	56	64	72
9	9	18	27	36	45	54	63	72	81
10	10	20	30	40	50	60	70	80	90

Students learn that multiplication and division can be used for a larger array of problems and begin to apply them in the various contexts.

Critical Area Two

In this critical area, students extend their understanding of fractions from adding, subtracting, and multiplying to include division of fractions. The goal is for students to be able to see the relationship between multiplication and division to understand why the method for dividing fractions works the way that it does. Students are then expected to take their newly established knowledge of fraction operations and apply these skills with all the other tools in their toolbox to solve problems.

Also in this critical area is the expectation that students will learn to extend their understanding of the number system to include all the whole numbers and their opposites, or what was formally known as the set of numbers called integers. Students work on developing an understanding about the order and absolute value of rational numbers (any real number of the form a/b, where a and b are integers, and b is not zero). The following Venn diagram shows the relationship between numbers in the real number system. As you can see, the rational number set includes all integers, whole numbers, and natural numbers.

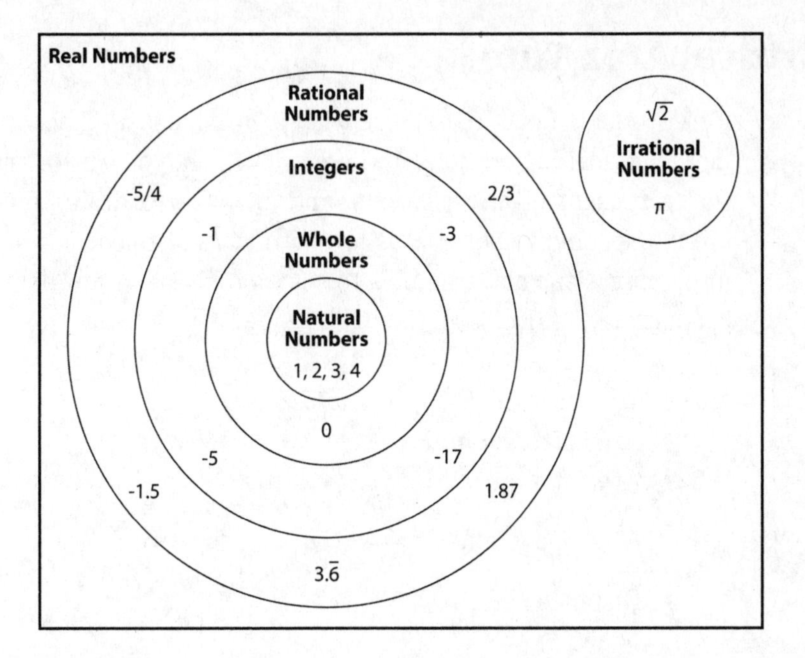

Students then take their understanding of rational numbers and apply it to the coordinate plane system, learning to plot ordered pairs in all four quadrants.

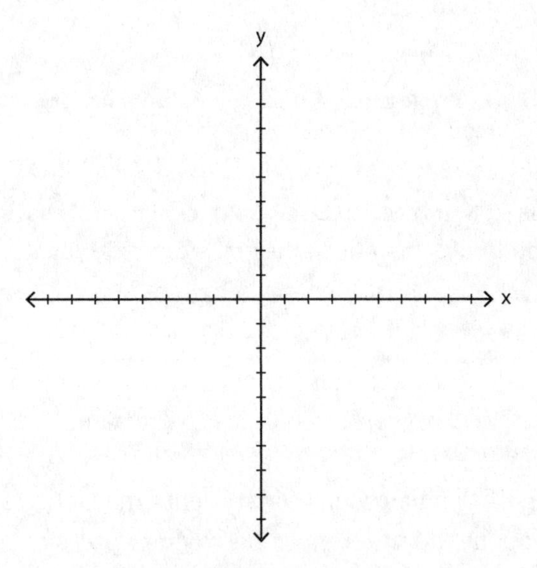

The coordinate plane system is created by the perpendicular intersection of two number lines, known as the x-axis (the horizontal number line) and the y-axis (the vertical number line). The point of intersection is called the *origin*. We will explore this in greater detail in Chapter 6.

Critical Area Three

Here students begin to explore what a variable is, and how they are used in mathematical expressions and equations. They learn to write expressions that represent given situations, and are able to evaluate the expressions given a specific value. At this level students also learn to substitute values into expressions and formulas to solve problems. A few examples of formulas that may be used are:

- Volume $= lwh = length \times width \times height$
- Area of a rectangle $= bh = base \times height$
- Area of a square $= s^2 = side \times side$
- Area of a triangle $= \frac{bh}{2} = (base \times height) \div 2$
- Distance $= rate \times time$

In this critical area, students also learn that although certain expressions may not appear equivalent, they in fact are. This can be proved by using the properties of operations to rewrite expressions.

Do you think these two expressions are equal?

$5(3x - 2)$ and $15x - 10$

The answer is yes, these two expressions are in fact equal. By applying the distributive property to the first expression you are able to prove this.

There are a lot more properties of operations that can be applied to prove that two expressions are equivalent. You'll learn more about that later on.

Another major focus for students in sixth grade is equations. Students learn about the differences between expressions and equations. They also learn that solutions to equations are values that, when substituted for a variable, make the equation true. Students also learn the beginning stages of solving one-step equations, understanding that when solving an equation, equality must be maintained.

The last aspect of this critical area is the idea of constructing and analyzing tables. Tables can be used to show equivalent ratios, or they can be used to describe relationships using equations such as $3x = y$. Creating tables to describe relationships between quantities is all about analyzing what happens between input and output values.

Critical Area Four

At this point, students should have a pretty good handle and solid understanding of numbers. With this in mind, students begin to explore and develop their ability to think statistically. Students will learn that a data distribution may not have a definite center, and that different ways to measure center provide different results. Students will learn about the role that mean, median, and mode play in analyzing data. The *median* measures center as the middle of the data values when organized from least to greatest. The *mean* measures center as it reflects the average of all the data points. The *mode* represents the data value that occurs most frequently in the data set. Lastly, students learn that a measure of variability (interquartile range, or mean absolute deviation) can be a useful tool for summarizing data. The goal of this set of standards is to help students learn a variety of ways to describe and summarize numerical data sets, identifying clusters, peaks, gaps, and symmetry, as well as thinking about how and where the data values were collected.

Additional Topics of Study

Although not a critical area, students in sixth grade spend time working on and developing their geometry skills that were first learned in elementary school. A focus on area, surface area, and volume occurs in sixth grade. Students learn to find areas of triangles and special quadrilaterals by breaking these shapes down and rearranging them, or by removing pieces, to make the more familiar shape of a rectangle (a shape for which they already know the formula to calculate area). Using these strategies, students are exploring and investigating ways to determine the area, and hopefully, to develop a formula all on their own.

Exploring the three-dimensional shapes of prisms and pyramids allows students to gain an understanding of surface area and volume. Once again, students break down three-dimensional shapes into pieces that they can relate to. Many times this means turning the three-dimensional prism into a collection of two-dimensional pieces that a student can determine the area of. In doing this, they are making connections between their understanding of two- and three-dimensional shapes.

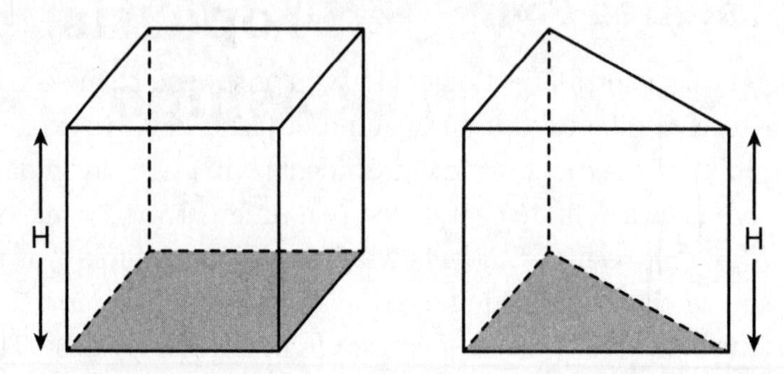

Students learn to calculate area, surface area, and volume by applying their knowledge of multiplying and dividing numbers with side lengths that are both whole number values as well as fractional values. Many students in sixth grade do well when these are whole numbers, but when the teacher introduces decimals or fractions for a side length, it can become difficult.

So there you have it, a quick and very brief outline of what is expected of your sixth grader. They will be busy all year, because along with all of this exciting math are a whole bunch of prerequisite skills that need to be built in, reviewed, and maintained for future years of study.

In the next few sections of the book, each of these areas will be explored in greater detail, and you will have a chance to practice your math skills with the practice problems included in each chapter. Get ready, because sixth grade is a jam-packed, fun-filled year of great mathematics!

CHAPTER 5

Ratios and Proportional Relationships

Sixth grade is the first year that children begin working with ratios and proportional relationships in a formal context. However, the skills leading up to this began in third grade, when students started developing their understanding of fractions as numbers. Further development continued in fourth grade, when students worked on fraction equivalence (recognizing two different fractions can have the same value). Lastly, fifth-grade students used fraction equivalence as a strategy to add and subtract fractions. This chapter will show you how the preliminary work with fractions leads to understanding of ratios and proportions.

What Should My Child Already Be Able to Do?

For students to be successful with ratios and proportions they need to have a set of prerequisite skills relating to fractions and whole number multiplication. Students should enter the sixth grade understanding how to multiply multi-digit whole numbers using the standard algorithm. Fluency with this skill requires students to know their multiplication facts to at least 100 (10×10). Knowing these two skills is essential to working with fractions.

Another important piece of knowledge to have before beginning work with ratios and proportions is to understand what makes two fractions equal. The acquisition of this skill begins in third grade when students learn about unit fractions ($\frac{1}{b}$, where b is any number other than 0). In this same year, students should be able to recognize and create simple equivalent fractions (for example, $\frac{1}{2} = \frac{2}{4}$ or $\frac{4}{6} = \frac{2}{3}$), and prove why this is true using a visual model. This very simple set of equivalent fractions will be their first experience with proportional reasoning, and they don't even know it yet.

In fourth grade, students focus on developing understanding of fraction equivalence with more difficult fractions, even improper fractions, like the example $\frac{15}{9} = \frac{5}{3}$, and work on developing methods for generating such pairs.

Lastly, your child needs to be able to multiply and divide decimals as well as fractions to answer questions related to the percent of a number. Decimal computation is a standard in the sixth grade, so hopefully your child's teacher has completed this unit of study prior to exploring percents. Fraction multiplication is taught in the fifth grade, but certainly is an area where students struggle, so a bit of review in this area wouldn't hurt.

If your child does not have these skills mastered, take time reviewing these skills at home. Consider purchasing a workbook at your local bookstore that is at a fourth or fifth grade level, and start there.

What Is a Ratio?

A ratio is a comparison of two quantities by division. It can be written in fraction form ($\frac{a}{b}$), with a colon (a:b), and in word form (a to b, or a out of b

where b cannot equal 0), depending on the type of ratio. There are two types of ratios that exist: a part-to-part ratio where you compare one part to another part (for example, girls to boys), or a part-to-whole ratio where you compare a part to a whole (for example, number of girls in a class to the total number of students). The word form "*a* out of *b*" can only be used with a part-to-whole ratio because the phrase "out of" implies a total amount of something.

In sixth grade, students are expected to be able to understand the meaning of a ratio and use it to describe the relationship between two quantities. If the words become confusing, encourage your child to draw a picture that represents the situation. This may help him move forward with the concept.

At the SPCA, there are 5 cats for every 3 dogs. Write this as a ratio comparing cats to dogs.

This is an example of a part-to-part ratio and would be written as $\frac{5}{3}$, 5 to 3, or 5:3, representing that there are 5 cats to 3 dogs.

The order of the numbers is important. Since the previous example asked the ratio to be written comparing cats to dogs, it is important that the number associated with cats is written first, followed by the number of dogs. If it were dogs to cats, the ratio would be written 3 to 5.

The way in which you write the ratio depends on the directions provided within the question. If it says a fraction, then it is important to use $\frac{a}{b}$. If it simply says write a ratio, then any of the three ways is perfectly acceptable.

Watch out for instructions that say, "write a ratio in simplest form." Students often write a correct ratio but miss the keywords in the direction that specify to simplify the ratio, and end up getting the problem incorrect.

Using Unit Rates

A rate is a special type of ratio that compares two quantities with different kinds of units. Think about all the different types of rates you already know and use in your life: heart rate, rate of speed, cost per amount of something, words read over a period of time, etc. A *unit rate* is a special type of rate. It expresses the rate for one unit of a given quantity. Here is a list of commonly used unit rates:

- Speed, miles per hour
- Heart rate, beats per minute
- Gas mileage, miles per gallon
- Gas price, cost per gallon
- Purchases, cost per item
- Reading/typing rate, words per minute

The list can go on and on, but you get the idea. Let's look at an example of purchasing chocolate fudge. If you pay $18.00 for 3 pounds of chocolate fudge, what is the unit rate?

$$\frac{\$18.00}{3\text{ pounds}} \div \frac{3}{3} = \frac{\$6.00}{1\text{ pound}}$$

This means that the unit rate is $6.00 per 1 pound of fudge.

When working with unit rates, it is always the denominator of the fraction that you want to simplify to 1. This is how you get the per-minute, per-pound, per-gallon rate. So consider this rule when setting up the ratio.

What is the unit rate of a 16-ounce jar of salsa that costs $2.88? Do you want to know how many ounces you can get for a dollar, or how many dollars you pay for an ounce? If you chose the second option, you would be correct. Even though the weight is given first, you want to set it up so that the cost is the numerator (top number of fraction) and the weight or amount is the denominator (bottom number of fraction).

$$\frac{\$2.88}{16 \text{ ounces}} \div \frac{16}{16} = \frac{\$0.18}{1 \text{ ounce}}$$

It will cost $0.18 per ounce of salsa.

Why does my child need to know about rates and unit rates?
It is important to understand how to use unit rates to become an educated consumer. Consider grocery shopping, for example. Do you price shop to get the best deal? If packages of products are sold in different sizes for different costs, how do you know which is the best deal? By using unit rates. Once you get the price per 1 pound, 1 ounce, 1 quart, and so on, of something, then you can compare the two costs. The lesser price per 1 quantity is the better deal.

Which Is the Better Buy?

A 12-pack of pencils sells for $4.20, and 15 pencils sell for $4.95. Which is the better buy?

Let's begin with the 12-pack. The rate to express the cost per quantity is $\frac{\$4.20}{12 \text{ pencils}}$. The unit rate for this is $0.35 per pencil, determined by dividing both the numerator and denominator by 12.

$$\frac{\$4.20}{12 \text{ pencils}} \div \frac{12}{12} = \frac{\$0.35}{1 \text{ pencil}}$$

The second rate is $\frac{\$4.95}{15 \text{ pencils}}$, which has a unit rate of $0.33 per pencil and is found by dividing both the numerator and denominator by 15.

$$\frac{\$4.95}{15 \text{ pencils}} \div \frac{15}{15} = \frac{\$0.33}{1 \text{ pencil}}$$

So which is the better buy? A pencil that costs $0.35 or one that costs $0.33? The one that costs $0.33. Two cents might not seem like a lot of savings, but when you start buying a lot of pencils, two cents saved per pencil adds up.

When working with money and determining unit rates, it is important to remember to round your answer to the nearest cent (nearest hundredth) because money only has two decimal places.

The idea of being an educated consumer is especially important these days when prices are skyrocketing and salaries are not. This skill is essential when trying to manage money as an adult. The good news is that many stores already have the unit price on the labels. It is typically in one of the upper corners in an orange or yellow rectangle. Pay attention to the unit of measure they are using.

Next time you are out shopping, show your child how unit prices are used at grocery stores and have him decide which product you should buy based on the unit price. It is a great real-world application of the math he is learning in school.

This basic level of ratio and rate understanding is just the beginning. Now let's use them to solve problems.

Applying Ratios and Rates to Solve Problems

At this stage, your child should have a firm grasp on what a ratio means. Being able to answer the basic questions is great, however, it is knowing how to apply them to more complex situations that makes learning about ratios and rates meaningful.

Using Unit Rates

Now that you know what a unit rate is, how can it be used to solve problems? If the know the cost for one unit, you can determine the cost for any amount of something, and vice versa. Let's say that you charged $25 dollars for 2.5 hours' worth of work; how much did you make an hour? Divide both the numerator and denominator by 2.5 and you get your hourly rate of $10.00 per hour. Now that you know this, you can solve any problem that is thrown at you.

What if the numbers are not so nice to work with? The same method still works. Divide both the numerator and denominator by the number in the denominator; this will simplify the fraction to a number over 1.

The Scale Factor Method

The scale factor method is exactly what the title means. Let's break the words down, starting with *scale*. Think about what it means to make a scale drawing of something. As defined by *www.mathisfun.com*, "a scale drawing is a drawing that shows a real object with accurate sizes except they have all been reduced or enlarged by a certain amount (called the scale)." Now let's look at the meaning of the word *factor*. Factors are numbers you multiply together to make a product. So a *scale factor* is a number you multiply by to make something larger or smaller. Let's put this into practice.

> If it takes you 4 hours to mow 3 lawns, how many lawns can be mowed in 28 hours? At what rate were lawns being mowed?

To begin this problem, you first want to set up a ratio representing the situation.

$$\frac{3 \text{ lawns}}{4 \text{ hours}}$$

If you continue at this rate, how many lawns can be mowed in 28 hours? To determine this, set the two ratios equal to one another and find what number you need to multiply the first fraction by to get to the second fraction.

$$\frac{3 \text{ lawns}}{4 \text{ hours}} = \frac{? \text{ lawns}}{28 \text{ hours}}$$

$4 \times 7 = 28$, so if you use a scale factor of $\frac{7}{7}$ we can figure out how many lawns you mowed.

$$\frac{3 \text{ lawns}}{4 \text{ hours}} \times \frac{7}{7} = \frac{21 \text{ lawns}}{28 \text{ hours}}$$

You are able to mow 21 lawns in 28 hours time.

To answer the last part of the question, you need to understand what the ratio means. If 3 lawns are mowed in 4 hours time, that means that $\frac{3}{4}$ of a lawn is mowed an hour, or that it takes 1 hour and 20 minutes to mow a lawn completely. Can you figure out how that was done?

A scale factor can also be used to simplify fractions to solve problems. Let's look at this example:

Mary can make 16 gallons of maple syrup in 4 days. At this rate, how many gallons of maple syrup can she make in 5 days?

16 gallons/4 days can be simplified by dividing by a scale factor of $\frac{4}{4}$.

$$\frac{16 \text{ gal}}{4 \text{ days}} \times \frac{4}{4} = \frac{4 \text{ gal}}{1 \text{ day}}$$

Now that we know the unit rate of 4 gal per day, we can use a scale factor of $\frac{5}{5}$ to answer the question.

$$\frac{4 \text{ gal}}{1 \text{ day}} \times \frac{5}{5} = \frac{20 \text{ gal}}{5 \text{ day}}$$

The answer to this problem is Mary can make 20 gallons of maple syrup in 5 days.

Using Ratio Tables

Oftentimes it is difficult for students to organize information when multiple steps are needed to solve a problem. This is where a table can come in handy. How many of you have ever made a drink from concentrate? This is a perfect example to use for a ratio table.

It takes 1 can of lemonade concentrate and 3 cans of water to make a batch of lemonade. If one batch serves 6 people, how much of each ingredient will be needed to serve 90 people? Use this information to complete the following table.

# of Batches	1		4			12			
# of Cans of Concentrate	1	3			10			20	
# of Cans of Water	3			21			45		162

In looking at the table, you will see there are three rows, labeled number of batches, number of cans of concentrate, and lastly, number of cans of water. Each column contains the amount needed for that many batches of lemonade. Using scale factors, you are able to complete the table.

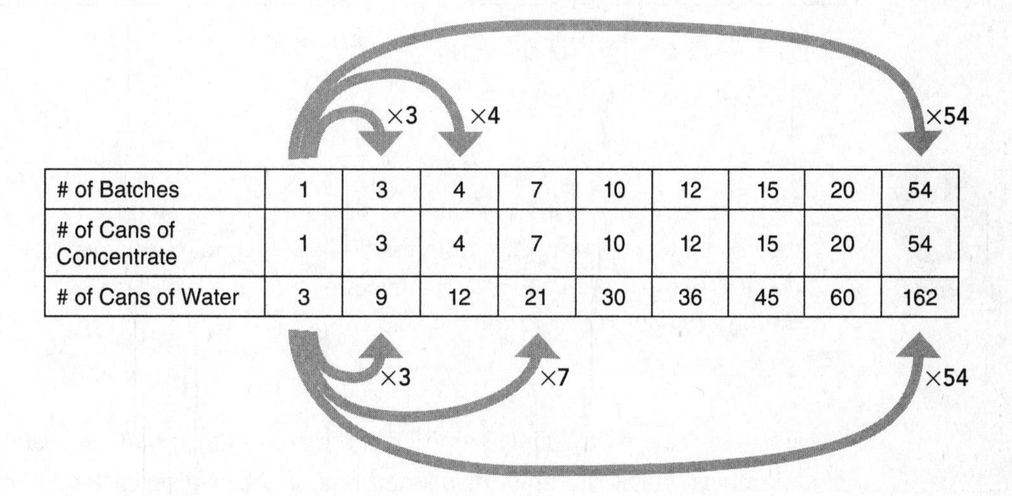

# of Batches	1	3	4	7	10	12	15	20	54
# of Cans of Concentrate	1	3	4	7	10	12	15	20	54
# of Cans of Water	3	9	12	21	30	36	45	60	162

Now the table is completely filled in. You answered the second part of the question that said to complete the table, but the first part asks how many batches are needed to serve 90 people. You can use the table to answer the question, but need to do a little legwork first.

Ratio tables can be set up horizontally or vertically. It is entirely up to the person solving the problem.

Using the unit rate of 6 people/1 batch and a scale factor of $\frac{15}{15}$, 90 people are served using 15 batches of lemonade. You can now reference the table and see that 15 batches requires 15 cans of lemonade concentrate and 45 cans of water.

Another method for solving ratio tables is to use addition. Since the ratio was 1 can of concentrate and 3 cans of water per batch, you can complete the table by adding 1 to the cans of concentrate, at the same time that you are adding 3 to the cans of water.

+1 +1 +1

# of Cans of Concentrate	1	2	3	4	5	6	7	8	9
# of Cans of Water	3	6	9	12	15	18	21	24	27

+3 +3 +3

A common mistake made by students using the addition method is that they add the same number to both the first and second row. Remember that it is a ratio, so you need to maintain the equivalence all the way through by adding on the original ratio.

Ratio tables are especially helpful if there are multiple questions about the same scenario, or if the amount needed changes based on different circumstances. Ratio tables also serve as a great way to present information to others.

Using Tape Diagrams

Tape diagrams are best used to model ratios when the two quantities have the same units. They are typically created using rectangular strips that are broken into equal-sized pieces. Each piece represents the same amount. Tape diagrams can be used to model part-to-part and part-to-whole ratio problems. This diagram is especially helpful for visual learners.

Let's look at a few examples using tape diagrams.

Part-to-Part Tape Diagram

Johnny likes to play a game called Astro-Bots. His current ratio of wins to losses is 3:2. If Johnny has won 15 games, how many games has Johnny lost?

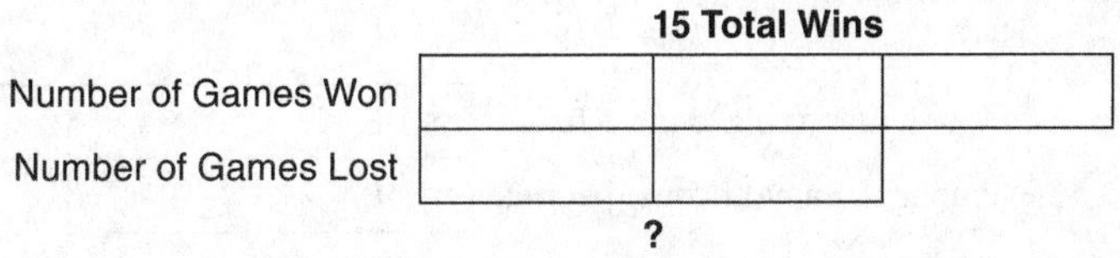

15 Total Wins

Number of Games Won			
Number of Games Lost			

?

The image here shows the initial setup of a tape diagram. We are given the information that Johnny has won 15 games. Notice this is labeled on the upper tape diagram. We do not know how many games Johnny lost, so a question mark is used to indicate the unknown. Given that Johnny has won 15 games, shared between three boxes, we know each box is worth 5 games.

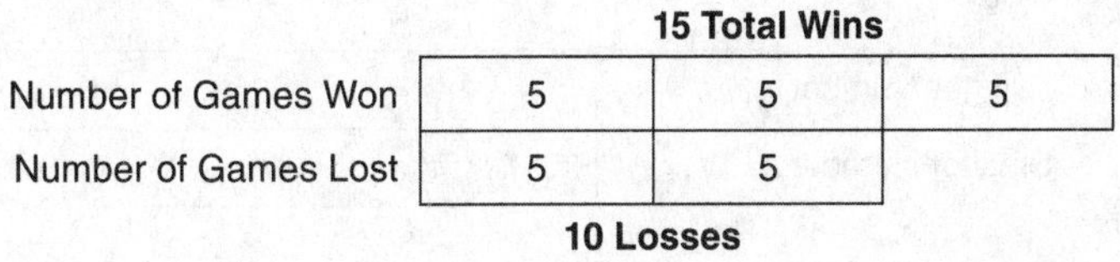

15 Total Wins

Number of Games Won	5	5	5
Number of Games Lost	5	5	

10 Losses

Based on the information provided, we were able to label each box as having a value of 5 games. Since the ratio is 3 to 2, that means that Johnny has lost 10 games. The tape diagram pictured supports this. This example is a part-to-part ratio model because it compares the part of the games won by Johnny to the part of the games lost.

Part-to-Whole Tape Diagram

The ratio of country tunes to easy-listening tunes on Derek's iPhone is 3:2. If Derek has 12 easy-listening tunes, how many tunes does he have in all? Let's build a model that represents the ratio in the word problem.

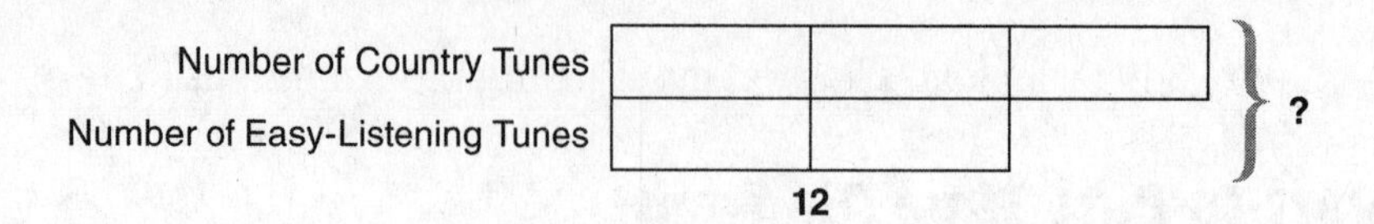

Notice the location of the question mark in this example. It includes all of the boxes, and is therefore representing the total amount of tunes Derek has on his iPhone. Based on the information that Derek has 12 easy-listening tunes, you can determine that each box is worth 6.

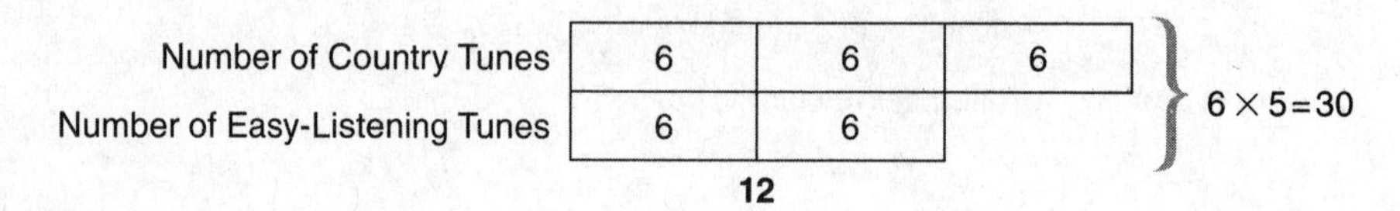

Knowing this allows you to determine that there is a total number of 30 tunes on Derek's iPhone.

Tape Diagram to Determine the Difference of Two Amounts

Two numbers are in the ratio of 7:4. If the smaller number is 24, what is the difference between the two numbers?

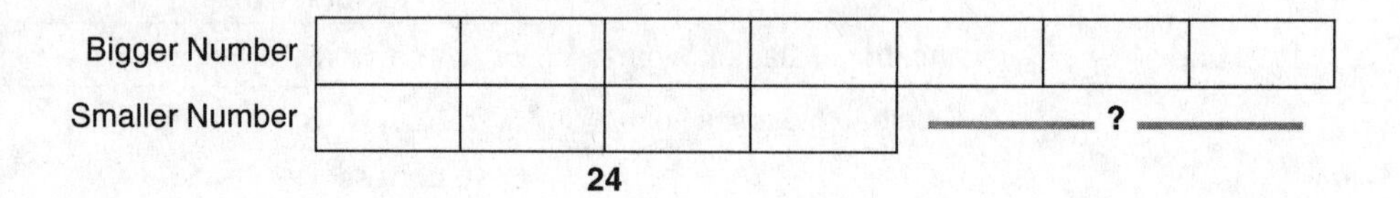

Since four boxes are worth 24, each box is worth 6. This allows you to continue to the second step of the tape diagram.

Bigger Number	6	6	6	6	6	6	6
Smaller Number	6	6	6	6		?	
		24					

Since the difference between the bigger number and the smaller number is three boxes, and each box is worth 6, the overall difference is worth 3×6, or 18.

Alternative Tape Diagrams

If your child has a difficult time drawing the boxes, consider providing him with hands-on materials to create the model. Blocks make a great manipulative because they are easy to move around and handle. Once your child has made the physical model, give him the opportunity to draw it on his own. If he is still struggling, you can act as a scribe, but ask him to dictate to you what it is he wants you to draw.

Lastly, *www.mathplayground.com* has an activity called Thinking Blocks, which can be downloaded as an app on your iPad or used in a web browser. You can access the Thinking Blocks for four different types of scenarios: Addition and Subtraction Word Problems, Multiplication and Division Word Problems, Fraction Word Problems, and Ratio and Proportion Word Problems. For this standard, please reference the Ratio and Proportion Word Problems and then pick from the various types of problems provided. What is great about this website is that it allows for students to practice the skill of modeling ratios using tape diagrams while giving instant feedback as to whether students are getting the answer right or wrong. It also has various levels of challenge, from beginner to advanced. Not only is this a great tool for your child, it is a great tool for you to use to develop an understanding of this standard, as well.

Using Double Number Lines

A double number line shows how the two units change in relation to one another. This is another diagram you can use to model ratios to solve problems. It is built on the same foundation of requiring the ratio to be added on, or by using multiplication with a scale factor. Let's look at a couple of examples to see how these take shape.

A party punch is made using 2 cups of Sprite for every 3 cups of Hawaiian Punch. If you already have 15 cups of Hawaiian Punch, how many cups of Sprite will you need to make the punch?

Students may look at this example and think of several different ways to solve it, ranging from drawing a picture, using a scale factor, creating a ratio table, or even drawing tape diagrams. However, there is an additional method that can be used and that is drawing a double number line diagram, which can combine many of the ideas previously discussed.

This method is exactly as it sounds. You begin by labeling the two items that are represented by the ratio: cups of Sprite and cups of Hawaiian Punch. Next, you draw two number lines that begin at 0 and extend to the right. If you have 0 cups of Sprite, you have 0 cups of Hawaiian Punch as well. Maintaining the ratio of 2 cups of Sprite for every 3 cups of Hawaiian Punch produces the following double number line diagram.

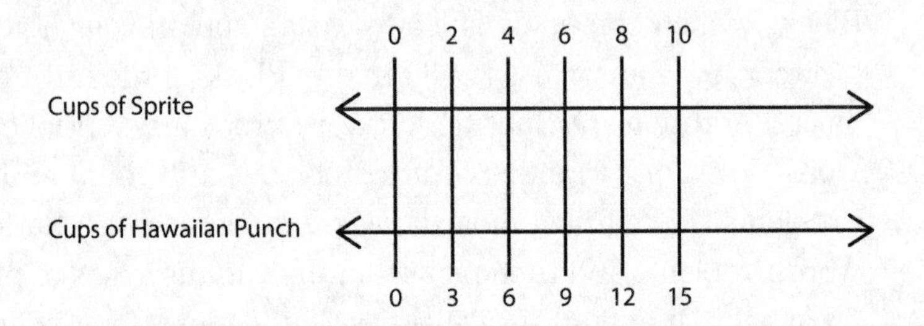

If you already have 15 cups of Hawaiian Punch, you will need 10 cups of Sprite. The same strategies that worked for the ratio tables work for the double number line diagrams, as well. You can choose to add on the ratio each time, or find a scale factor to multiply both parts of the ratio by to find an equivalent ratio that answers the question.

Here is a second example to review to understand how to create your own double number line. A car travels at a rate of 65 miles per hour. How far does the car travel in 3 hours' time?

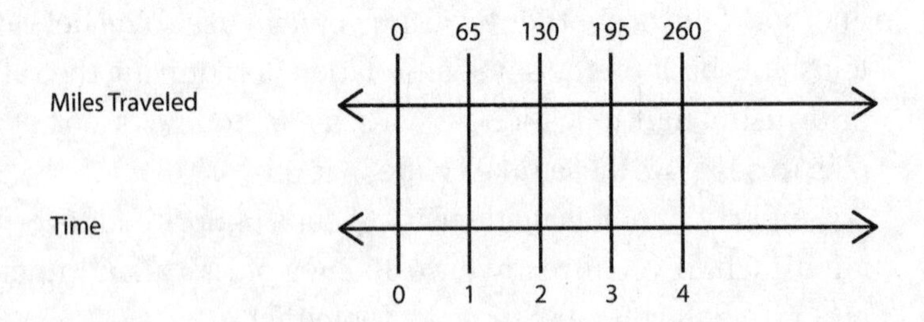

A car can travel 195 miles in 3 hours, traveling at an average speed of 65 miles per hour.

You will have a chance at the end of this chapter to practice these skills on your own. Get ready, as the challenge awaits you!

Finding a Percent of a Number

Let's begin with the most basic idea of a percent. A percent is a ratio out of 100. It compares part of a whole. For example, 5% is written as a ratio of 5 out of 100, or as a fraction: $\frac{5}{100}$. If you were to draw a model of 5%, it would look like this:

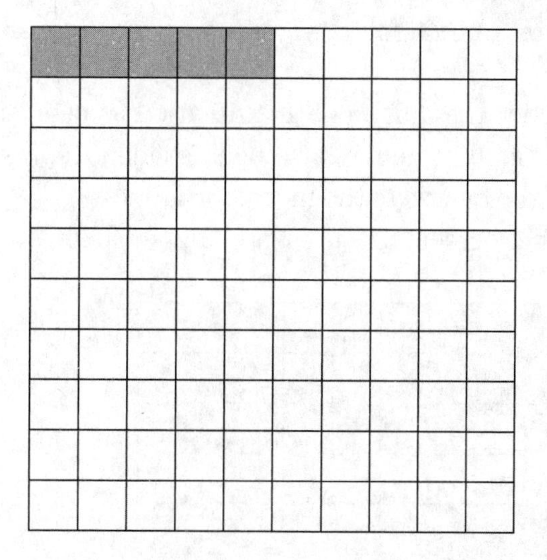

Notice that this is a 10×10 grid that has 100 squares. 5 of those squares have been shaded, modeling 5% or the ratio $\frac{5}{100}$. This model can be used to represent any percent, and is a great tool to help students visualize a part out of 100.

What would happen if you were asked to find 110% of something? What would this look like as a fraction? Since percents are out of 100, $110\% = \frac{110}{100}$. This is an improper fraction. Let's simplify it. How many times does 100 go into 110? It goes 1 time, with 10 left over. This is written as $110\% = \frac{110}{100} = 1\frac{10}{100}$. What does the model look like?

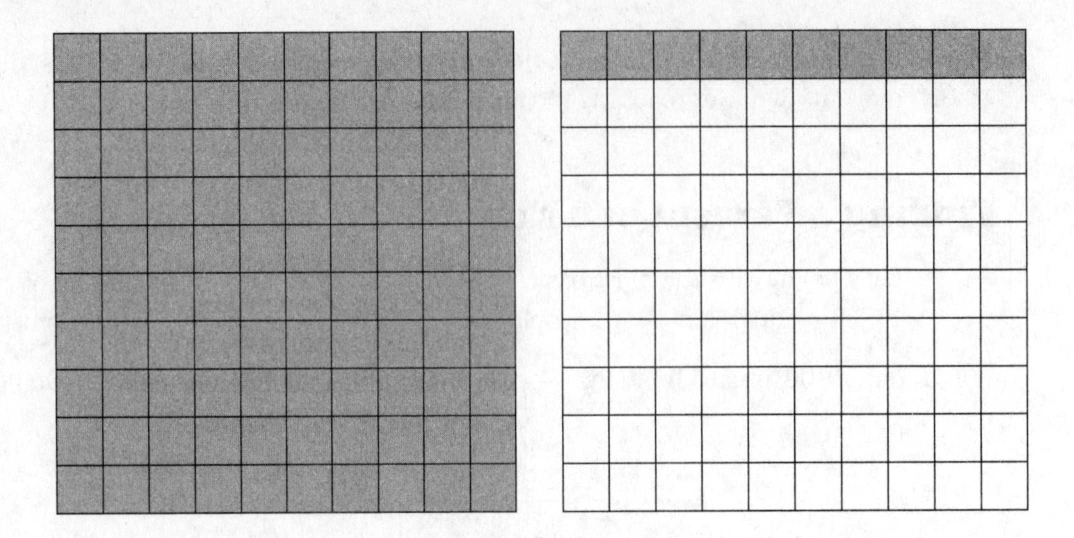

Notice that all of one 10 × 10 grid has been shaded, showing 100%, and 10 out of 100 squares have been shaded, representing 10%. Putting them together, you create the model for 110%.

What percent is represented by the following model?

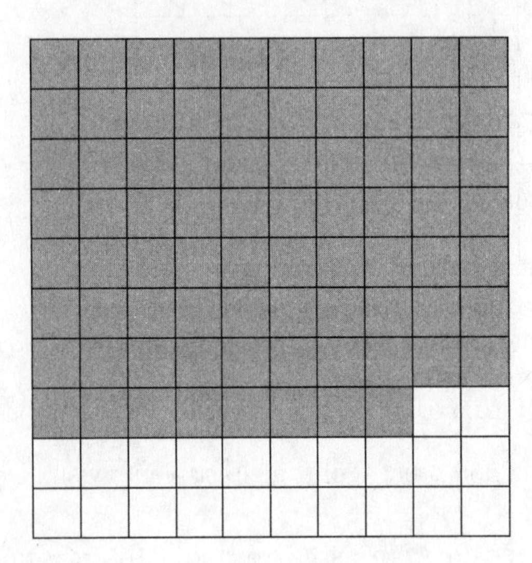

If you said 78%, you would be correct. 78 out of the 100 squares are shaded.

It is easy to write a whole number percent as a fraction, but be careful if it includes a decimal. Remember, the first step is to write the percent over 100, but you cannot leave a decimal in a fraction at this point in time. Using a power of ten as a scale factor will change the numerator to a whole number.

$$3.5\% = \frac{3.5}{100} \times \frac{10}{10} = \frac{35}{1000}$$

$$0.27\% = \frac{0.27}{100} \times \frac{100}{100} = \frac{27}{10{,}000}$$

Write Percents as Fractions and Decimals

Being able to switch between forms of a number will help your child become more proficient when working with percents. Percents can be written as a percent, as a fraction, or as a decimal. The form you use often depends on the numbers being used in the question.

Using 75% to get us started, let's write this in the different forms.

75% as a decimal $= \frac{75}{100} = 0.75$ (This decimal is read as seventy-five hundredths. Percents work really nicely with our base ten number system.)

75% as a fraction $= \frac{75}{100} \div \frac{25}{25} = \frac{3}{4}$ (The ratio $\frac{75}{100}$ has been simplified to lowest terms, and is equal to $\frac{3}{4}$.)

What about 140%?

140% as a decimal $= \frac{140}{100} = 1\frac{40}{100} = 1.40$ (This decimal is read as 1 and forty hundredths.)

140% as a fraction $= \frac{140}{100} = 1\frac{40}{100} \div \frac{20}{20} = 1\frac{2}{5}$

Some percents expressed as fractions can be simplified, but not all. Remember, if the question asks you to write the percent as a fraction, it should always be in its simplest form.

Now that you know how to write percents as fractions and as decimals, you can move on to the next step.

Finding the Percent of a Number

Remember that a percent represents a part out of a whole. If you are looking for a percent of a number, you are looking for that part of another number. This section will show you a variety of strategies that can help you and your child find a percent of a number.

Using Benchmark Percents

A benchmark percent is a commonly used percent that has an easy fraction equivalent. Here is list of the most commonly used benchmark percents:

- $1\% = \frac{1}{100}$ of a number
- $5\% = \frac{5}{100} = \frac{1}{20}$ of a number
- $10\% = \frac{10}{100} = \frac{1}{10}$ of a number
- $25\% = \frac{25}{100} = \frac{1}{4}$ of a number
- $33\frac{1}{3}\% = \frac{33.33333}{100} = \frac{1}{3}$ of a number
- $50\% = \frac{50}{100} = \frac{1}{2}$ of a number
- $66\frac{2}{3}\% = \frac{66.666666}{100} = \frac{2}{3}$ of a number
- $75\% = \frac{75}{100} = \frac{3}{4}$ of a number
- $100\% = \frac{100}{100} =$ all of a number

Help your child memorize these benchmark percents, because knowing them will help him use mental math skills to do quick calculations of a percent of a number. In fact, any percent can be found using a combination of the benchmark percents listed.

27% of 40 can be found using the following combination.

$25\% + 1\% + 1\%$

25% of $40 = \frac{1}{4}$ of $40 = 10$

1% of $40 = \frac{1}{100}$ of $40 = 0.4$, so 2% of $40 = 2 \times 0.4 = 0.8$

$25\% + 2\% = 27\%$

$10 + 0.8 = 10.8$, so 27% of 40 is equal to 10.8

Let's give this another try. What is 35% of 80?

35% can be made up of the following benchmark percents:

$10\% + 10\% + 10\% + 5\%$

10% of $80 = 8$ (You can quickly calculate this in your head by dividing 80 by 10.)

30% of $80 = 3(10\%$ of $80) = 3(8) = 24$

5% of $80 = 4$ (or $\frac{1}{2}$ of 10% of 80)

Put this all together and you get $24 + 4 = 28$, so 35% of $80 = 28$. No calculators, just simple mental math calculations. Using mental math to calculate percents requires fluency with a fractional part of a number. You can help your child improve his fluency by asking him to do the mental math calculations when you are out shopping and there are sales, or when leaving a tip at a restaurant. As a parent, you are pulling the math that students are learning into the real world that surrounds them!

Using Fraction Multiplication

Your child can write a percent as a fraction, and she learned in fifth grade to multiply fractions, so let's pull these two ideas together and allow your child to find a percent of a number using fraction multiplication.

To multiply fractions, multiply numerator times numerator, and denominator times denominator. Simplify the resulting fraction.

What is 8% of 15?

Let's walk through it step by step.

1. **Step 1:** Write the percent as a fraction.

   $8\% = \frac{8}{100}$

2. **Step 2:** Write the whole number as a fraction.

   $15 = \frac{15}{1}$

3. **Step 3:** Set up the multiplication problem.

   $\frac{8}{100} \times \frac{15}{1}$

4. **Step 4:** Solve by multiplying the two numerators (top numbers) and the two denominators (bottom numbers).

   $\frac{8}{100} \times \frac{15}{1} = \frac{120}{100}$

5. **Step 5:** Simplify the fraction and write as a number (may be a decimal).

   $\frac{120}{100} = 1\frac{20}{100} = 1.20 = 1.2$ (the zero at the end of the number is not necessary).

Now, that wasn't so bad, was it?

Decimal Multiplication

Decimal multiplication is a great strategy to use for percent-of-a-number problems that are often harder to do because they do not include nice, user-friendly numbers. In this strategy, you convert the percent to a decimal and multiply it by the number you are trying to find the part of. Again, let's break it down into step-by-step directions.

What is 4.5% of 25.8?

1. **Step 1:** Write 4.5% as a decimal.

   $4.5\% = \frac{4.5}{100} = \frac{45}{1000} = 0.045$

2. **Step 2:** Set up the decimal multiplication problem. Always write the number with more place values first. In this case they are equal so it doesn't matter.

$$\begin{array}{r} 0.045 \\ \times\ 25.8 \\ \hline \end{array}$$

3. **Step 3:** Solve using traditional multiplication.

$$\begin{array}{r} 0.045 \\ \times\ 25.8 \\ \hline 360 \\ 2250 \\ +9000 \\ \hline 1.1610 \end{array}$$

The answer to 4.5% of 25.8 is 1.161.

Using a Double Number Line

For this example, you'll need a skill previously learned with ratios to find a percent of a number. Let's use a double number line to determine 60% of 120.

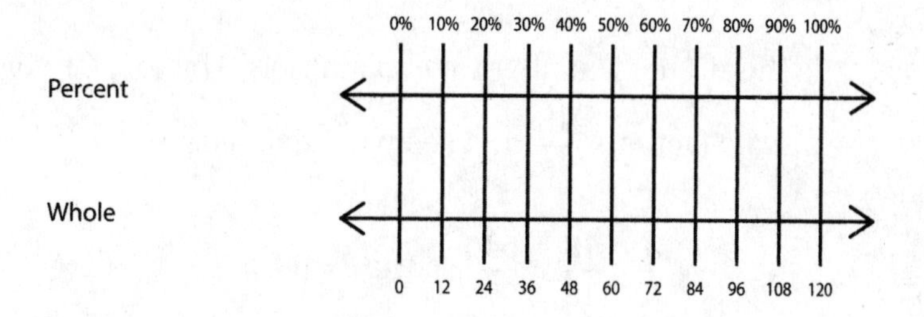

The double number line illustrates that 60% of 120 is 72.

Find the Whole from a Percent

This is the second type of percent problem your child will learn in sixth grade. In these examples, you can use equivalent ratios to find the whole, when given a part and the percent. We use what is called the *percent proportion* (a proportion is two ratios set equal to one another) to find this:

$$\frac{\text{percent}}{100} = \frac{\text{is part}}{\text{of whole}}$$

Let's try an example. 54 is 60% of what number? To help students be successful with these types of problems, encourage them to use different symbols to identify each part of the problem.

54 is 60% of what number?

$$\frac{\%}{100} = \frac{\textit{is part}}{\textit{of whole}}$$

$$\frac{60}{100} = \frac{54}{x}$$

Now, that still doesn't look easy to solve, but if we simplify the $\frac{60}{100}$ by dividing by $\frac{20}{20}$, we get the fraction of $\frac{3}{5}$. Let's try and work with this . . .

$$\frac{3}{5} = \frac{54}{?}$$

Is there a nice number we can multiply 3 by to get to 54? Yes . . . 18. Let's use a scale factor of $\frac{18}{18}$ and see what happens.

$\frac{3}{5} \times \frac{18}{18} = \frac{54}{90}$, so the ? = 90, and 54 is 60% of 90.

You did it. You have used the strategies we have explored to solve the problem.

Measurement Conversions

In this section, you will use what you have learned about ratio reasoning to convert measurement units. To convert a unit of measure, you need to multiply or divide by a conversion factor. A *conversion factor* is a rate in which the two quantities are equal, but are expressed using different units.

Convert Units of Length

To convert between different units of length, you need to know the conversion factors. Many people do not have these memorized and need to look them up. However, memorizing a few of the most common conversion factors will make your life and the life of your child a lot easier. Here is a table with a few of the most commonly used customary units of length.

▼ CUSTOMARY UNITS OF LENGTH

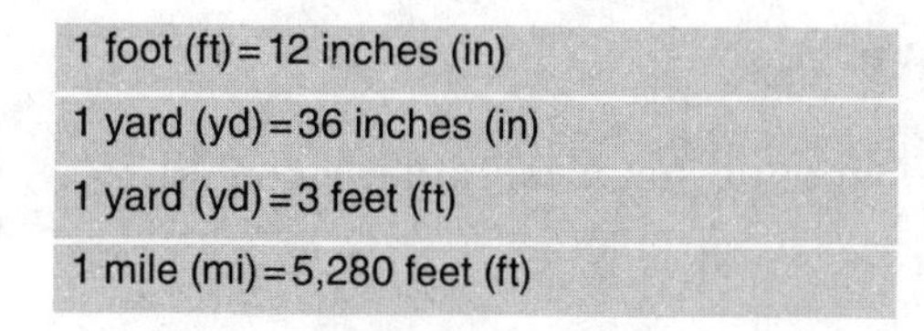

1 foot (ft) = 12 inches (in)
1 yard (yd) = 36 inches (in)
1 yard (yd) = 3 feet (ft)
1 mile (mi) = 5,280 feet (ft)

Now, let's complete some conversions.
Convert to the given unit.

$$3\ \text{feet} = \underline{\qquad\qquad}\ \text{inches}$$

1. **Step 1:** Choose a conversion factor.

   1 foot = 12 inches, so use the conversion factor of $\frac{12\ \text{inches}}{1\ \text{foot}}$.

2. **Step 2:** Multiply the given by the conversion factor.

   $\frac{3\ \text{feet}}{1} \times \frac{12\ \text{inches}}{1\ \text{foot}}$

3. **Step 3:** Simplify and cancel out the units of measure.

   $\frac{3\ \cancel{\text{feet}}}{1} \times \frac{12\ \text{inches}}{1\ \cancel{\text{foot}}} = 36\ \text{inches}$

So, 3 feet = 36 inches

What if you need to go from a bigger unit to a smaller customary unit; what would you do? The answer is: use division.

$$13{,}200\ \text{feet} = \underline{\qquad}\ \text{miles}$$

1. **Step 1:** Choose a conversion factor.

   1 mile = 5,280 feet

2. **Step 2:** Multiply by the conversion factor (the placement in the denominator actually means division).

$$\frac{13{,}200\text{ ft}}{1} \times \frac{1\text{ mi}}{5{,}280\text{ ft}}$$

3. **Step 3:** Simplify and cancel out the units of measure.

$$\frac{13{,}200\,\cancel{\text{ft}}}{1} \times \frac{1\text{ mi}}{5{,}280\,\cancel{\text{ft}}} = \frac{13{,}200}{5{,}280\text{ miles}}$$, which simplifies to 2.5 miles

Convert Units of Capacity

Capacity measures the amount that a container can hold. Here is a reference table to use to help solve the problems.

▼ **CUSTOMARY UNITS OF CAPACITY**

8 fluid ounces (fl oz) = 1 cup (c)
2 cups (c) = 1 pint (pt)
2 pints (pt) = 1 quart (qt)
4 cups (c) = 1 quart (qt)
4 quarts (qt) = 1 gallon (g)

Following the steps will allow you to complete the conversions. Convert to the given unit.

35 cups = _____ quarts

If 1 quart = 4 cups, use this as the conversion factor.

$$\frac{35\,\cancel{\text{cups}}}{1} \times \frac{1\text{ quart}}{4\,\cancel{\text{cups}}} = \frac{35\text{ quarts}}{4} = 8\frac{3}{4}\text{qt.}$$

You can rename the fractional part using the smaller unit.

$$\frac{35\text{ quarts}}{4} = 8\frac{3}{4}\text{qt.}$$

Convert Units of Weight and Mass

▼ **CUSTOMARY UNITS OF WEIGHT**

1 pound (lb) = 16 ounces (oz)
1 ton (T) = 2,000 pounds (lb)

Convert to the given unit.

19 pounds = _____ ounces

16 ounces = 1 pound is the conversion factor.

$$\frac{19\cancel{\text{ lb}}}{1} \times \frac{16 \text{ oz}}{1\cancel{\text{ lb}}} = \frac{304 \text{ oz}}{1}$$

So, 19 pounds = 304 ounces

Converting Metric Units

The metric system is the internationally agreed-upon system of measurement. Students primarily learn this system in their science classes. However, it still requires the use of math to go from one unit to another.

To aid students in learning the different prefixes, a metric staircase was developed. This helps students move between prefixes easily, because it is all based on multiplying and dividing by a power of 10.

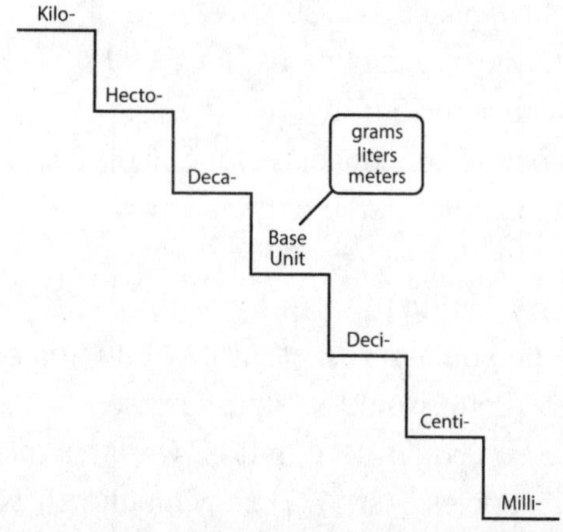

When converting from one prefix to another, count the number of steps from the first prefix to the second and then move your decimal point that many places in the same direction.

The mathematics that allows movement of the decimal point is multiplying or dividing by a power of 10. If you are moving the decimal point to the right, you are multiplying by a power of 10. Movement to the right 2 spaces happens by multiplying by 10^2. If you are moving the decimal point to the left, you are dividing by a power of 10. Movement to the left 3 spaces happens by dividing by 10^3.

Students also develop mnemonic devices to help remember the prefixes. The order of prefixes from large to small in the metric system is Kilo-, Hecto-, Deca-, Unit, Deci-, Centi-, Milli-. Students can develop their own mnemonic device or use one that is already developed like, King Henry Does Usually Drink Chocolate Milk.

Practice Problems

Write a ratio in simplest form to represent each scenario. Identify this as a part-to-part ratio or a part-to-whole ratio.

1. There are 5 girls for every 3 boys in class.
2. There are 8 boys in a class of 20 students.
3. There are 9 peaches and 12 apples.
4. There are 6 chairs for every table.

Calculate the unit rate.

5. 24 cupcakes for $26.88
6. 16-ounce box of cereal for $3.99
7. 5 miles in 55 minutes
8. 8 pounds of chocolate for $36.00
9. 1,652 miles in 3.5 hours

Use ratio reasoning to solve word problems.

10. A person can run at a rate of 20 miles in 120 minutes. At that rate, how long would it take a person to run 15 miles? How far could a person run in 12 minutes? Create a ratio table to solve this problem.
11. If 6 apples cost $1.50, how much will 24 apples cost?
12. Out of 120 students surveyed at the Cooperative Middle School, 30 said they buy a hot lunch. Based on these results, predict how many

of the 1,350 students at the Cooperative Middle School buy a hot lunch.

13. Noah had $\frac{1}{7}$ as many sweatshirts to sell as Jada. The two friends decided to share the work evenly. To do that, Jada had to give 21 sweatshirts to Noah. What was the total number of sweatshirts Jada needed to sell by herself before sharing the work with Noah?
14. For every 5 crunches that Kyle did during the physical fitness test, Caleb did 3. Kyle was able to do 14 more crunches than Caleb. How many crunches did Caleb do?

Calculate the percent of the quantity using any strategy.

15. 60% of 50 is ______.
16. 10% of 92 is ______.
17. 125% of 72 is ______.
18. 0.5% of 500 is ______.
19. Fred correctly answered 95% of the 40 problems on his math test. How many questions did he answer correctly?

Find the unknown value.

20. 15 is 20% of what number?
21. 42 is 50% of what number?
22. 12 is 120% of what number?
23. 27 is 30% of what number?

Convert between each of these units of measurement.

24. $8\frac{1}{2}$ gallons = ________ quarts
25. 25 feet = ________ inches
26. 150 seconds = ________ minutes
27. $2\frac{1}{4}$ tons = ________ pounds
28. $3\frac{1}{3}$ yards = ________ feet
29. 12 pounds = ________ ounces

CHAPTER 6

The Number System

This chapter is packed full of the prerequisite skills that students will need to be successful in seventh and eighth grade. It is here that students learn the last operation of fractions: division. Students will be able to compute quotients of fractions, solve word problems involving division of fractions, and draw visual models to represent a problem. The ability to fluently compute any type of fraction operation will be key in their future success. Along with division of fractions, students will be able to perform computation of multidigit numbers with mastery. This includes addition, subtraction, multiplication, and division, all using the traditional algorithm. This chapter will delve into all these areas and show you what you need to know.

What Should My Child Already Be Able to Do?

Students come from elementary school with a wide variety of skills. There, they learn the fundamentals for success in middle school. Students in fifth grade are expected to be able to "perform operations with multidigit whole numbers and with decimals to the hundredths," as stated by Corestandards.org. This includes multiplication of multidigit numbers using the standard algorithm, dividing a four-digit number by a two-digit number, and adding, subtracting, multiplying, and dividing decimals to the hundredths, using models or drawings. What this means is that students should have some understanding of these concepts, and, although they have not mastered all of them yet using the traditional algorithm, a certain level of understanding and proficiency should be evident.

Also, fifth-grade students should have learned to multiply fractions in a very basic sense, multiplying a fraction times a whole number or a fraction times a fraction. Students' understanding of division of fractions is limited to dividing unit fractions by whole numbers, and whole numbers by unit fractions.

Once again, knowledge of multiplication math facts is crucial when we work with dividing fractions, dividing using the standard algorithm, and calculating the greatest common factors and least common multiples of two numbers.

Dividing Fractions

Here we go, back into the dreaded world of fractions—at least, that's how many people feel. For whatever reason, fraction operations is the one concept that students always struggle with, and dividing fractions doesn't get any easier for them. This is one of the areas that experienced a lot of changes in the Common Core. The standard no longer calls for students to simply be able to compute quotients of fractions; it now requires students to use visual fraction models to represent the problem. This is where the conceptual understanding is developed. This is the "new math," but don't let this scare you. The idea of having students drawing a picture to make meaning of what it is to divide a whole number by a fraction, or divide a fraction by a fraction, gives them another way to think about mathematics.

Let's begin with a very basic problem. If you have 3 cups of flour and need $\frac{3}{4}$ cup of flour to make 1 batch of chocolate chip cookies, how many batches of cookies can you make? We begin by writing out the equation to represent the problem. $3 \div \frac{3}{4} = ?$ Next, let's create the visual model.

Notice that all that is drawn is the initial fraction, because it shows the amount we have to start with. Our next step is to break each whole into 4 parts. Why? Because it is the denominator of the second fraction.

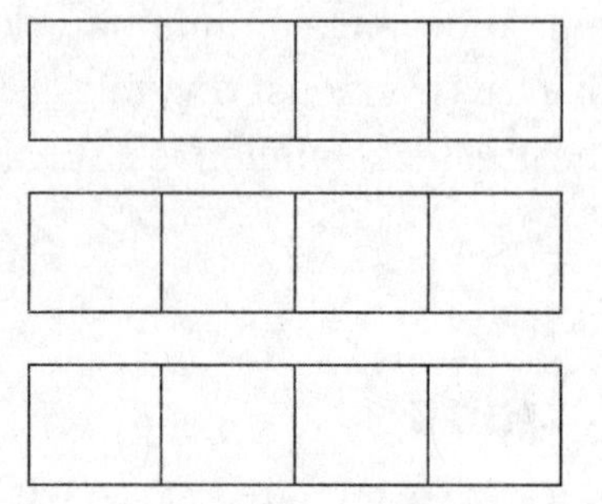

And now, using the model, determine how many times $\frac{3}{4}$ fits into 3. This is done by shading three out of every four parts.

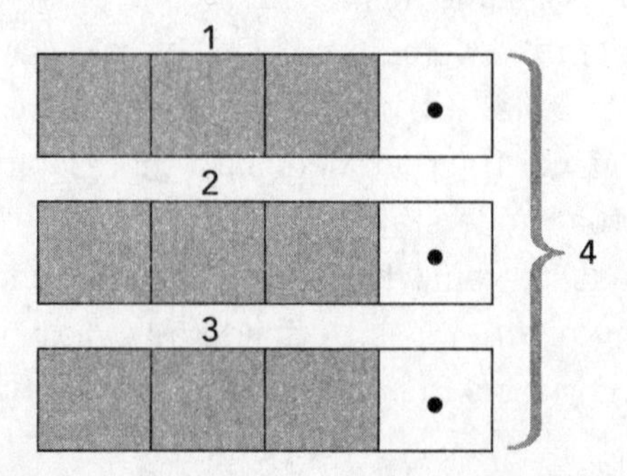

Using the model, we see that $3 \div \frac{3}{4} = 4$, so 4 batches of chocolate chip cookies can be made using 3 cups of flour, but now let's prove it using an algorithm. Here's a little tip taken from *www.mathisfun.com*: "Dividing fractions is easy as pie, flip the second and multiply." This little rhyme will help your children remember what they need to do to divide fractions.

$$3 \div \frac{3}{4} = 3 \bullet \frac{4}{3} = \frac{12}{3} = 4 \text{ batches}$$

That wasn't so hard.

Let's look at another example, this time a fraction divided by a fraction.

$\frac{3}{4} \div \frac{3}{8} = ?$

Again, we begin by creating a model to represent what it is we have, $\frac{3}{4}$.

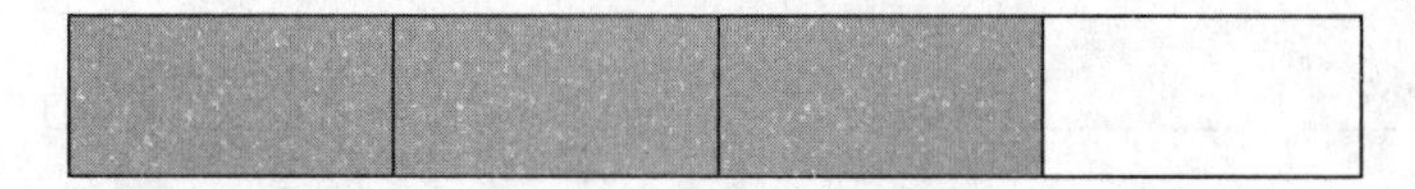

Now we need to break this fraction into eighths.

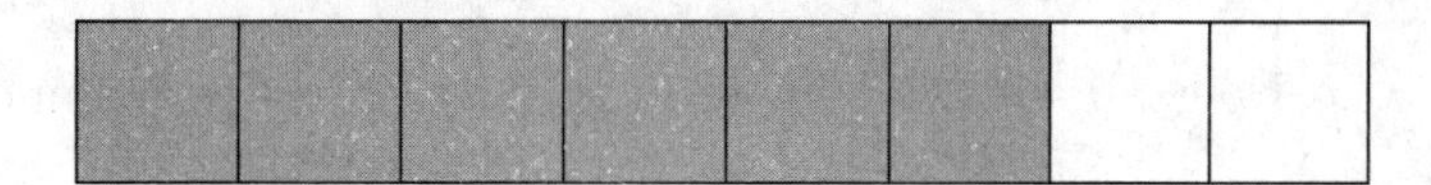

How many times does $\frac{3}{8}$ fit into $\frac{3}{4}$? Look at the diagram, what do you think?

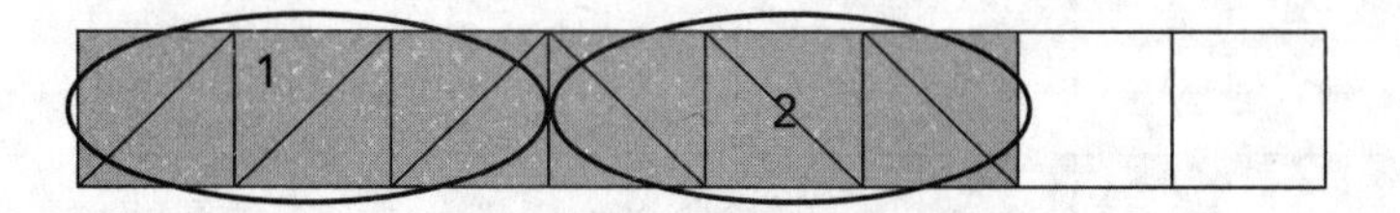

$\frac{3}{8}$ fits into $\frac{3}{4}$ two times, so $\frac{3}{4} \div \frac{3}{8} = 2$. Again, this one still works out nicely. What about when they are not so nice?

How many $\frac{3}{4}$-cup servings are in $\frac{2}{3}$ of a cup of ice cream? Here, we have $\frac{2}{3}$ cup of ice cream and we want to know how many $\frac{3}{4}$-cup servings there are. The math problem to represent this is $\frac{2}{3} \div \frac{3}{4}$. Is $\frac{3}{4}$ more or less than $\frac{2}{3}$? It is more. Will the quotient be greater than or less than 1? Less than 1. Using mathematical reasoning to predict the value of your answer is a very helpful skill. It allows your child to answer the question, "Does my answer make sense?"

This problem is where it gets tricky. We are going to build a model for both of these fractions, answer some questions, and then try to develop a solution.

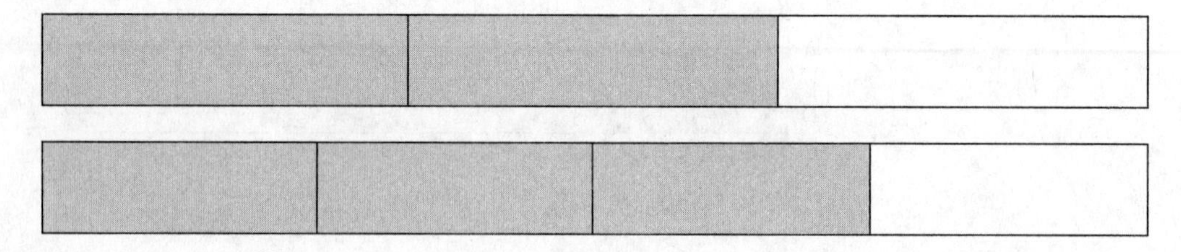

Does all of the $\frac{3}{4}$ fit into the $\frac{2}{3}$? No. Well, how much of it does? Let's break both fractions into equal-sized pieces. What is the least common multiple of 4 and 3? It's 12. Both 3 and 4 go into 12 evenly. If we break each of the fractions into twelfths, we can divide them.

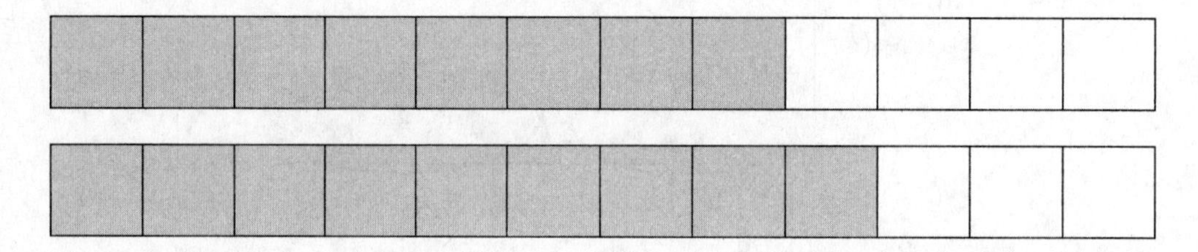

Notice that equivalent fractions have been created. We are now showing that $\frac{2}{3} = \frac{8}{12}$ and that $\frac{3}{4} = \frac{9}{12}$, and the new division problem is $\frac{8}{12} \div \frac{9}{12}$.

How many twelfths would we need for $\frac{9}{12}$ to go in one time? We would need nine twelfths. We only have eight, so $\frac{8}{9}$ of $\frac{9}{12}$ fits into $\frac{8}{12}$, or $\frac{8}{9}$ of $\frac{3}{4}$ fits into $\frac{2}{3}$.

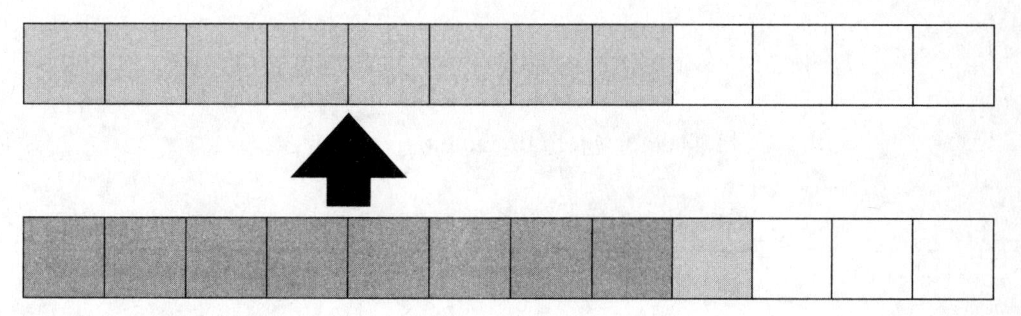

8 out of the 9 pieces fit into $\frac{2}{3}$

$\frac{2}{3} \div \frac{3}{4} = \frac{2}{3} \bullet \frac{4}{3} = \frac{2 \bullet 4}{3 \bullet 3} = \frac{8}{9}$

Hmm, let's do the algorithm and see if this works out.

Here are the steps to dividing fractions.

1. Step 1: Make sure all fractions are in fraction form. This means change all mixed numbers or whole numbers into improper fractions.
2. Step 2: Change division to multiplication.
3. Step 3: Take the *reciprocal* of the second fraction (this means "flip" the second fraction).
4. Step 4: Multiply the numerators, multiply the denominators.
5. Step 5: Simplify.

$$\frac{2}{3} \div \frac{3}{4} = \frac{2}{3} \bullet \frac{4}{3} = \frac{2 \bullet 4}{3 \bullet 3} = \frac{8}{9}$$

It works!

Continue to explore with models of different fraction problems. Create your own division problems for your child and see if he can solve them. Start with easier problems and work your way up to more challenging options.

The final piece is to show you how to divide two mixed numbers.

$$4\frac{1}{2} \div 1\frac{3}{4} = ?$$

1. **Step 1:** Write each mixed number as a fraction. Do this by multiplying the denominator times the whole number, and adding it to the numerator, keeping the denominator the same.

   $2 \bullet 4 + 1 = 9$, so $4\frac{1}{2} = \frac{9}{2}$ $\qquad$ $4 \bullet 1 + 3 = \frac{7}{4}$

   And the problem becomes $\frac{9}{2} \div \frac{7}{4} = ?$

2. **Step 2:** Change division to multiplication.

   $\frac{9}{2} \bullet \underline{\qquad} = ?$

3. **Step 3:** Multiply by the reciprocal of the second fraction.

   $\frac{9}{2} \bullet \frac{4}{7} = ?$

4. **Step 4:** Multiply the fractions.

   $\frac{9}{2} \div \frac{7}{4} = \frac{9}{2} \bullet \frac{4}{7} = \frac{9 \bullet 4}{2 \bullet 7} = \frac{36}{14} = 2\frac{8}{14}$

5. **Step 5:** Simplify.

   $\frac{36}{14} = 2\frac{8}{14} \div \frac{2}{2} = 2\frac{4}{7}$

Students often make mistakes when dividing fractions. They remember to change the division symbol to multiplication, but often forget to flip the second fraction. Another mistake is they flip the second fraction, but forget to change the operation. Either way, things will not work out correctly.

Phew! That was a lot of information, and we really only skimmed the surface of fraction division. There are so many levels and different ways to simplify that an entire chapter could be spent on just that operation. Hopefully, you've at least gained perspective on what it means to model fraction division. Your child's textbook will have more examples on ways to divide fractions, including the various ways to simplify before you multiply.

Dividing Using the Standard Algorithm

Let's make some connections to what we already know. We know the traditional algorithm for multiplication, and we also know that multiplication is nothing more then repeated addition. $2+2+2=2\bullet3=6$. Two was repeatedly added three times, so the multiplication problem is $2\bullet3$.

So if division is the inverse (opposite) operation to multiplication, what is division? You guessed it: *repeated subtraction*. For smaller numbers, this idea of repeated subtraction is easy.

$$24\div6=24-6-6-6-6=0$$

Six was repeatedly subtracted from 24 four times, so $24\div6=4$.

If we begin to use larger numbers it becomes more of a daunting task.

$$226\div25$$

Let's use repeated subtraction to figure out this problem.

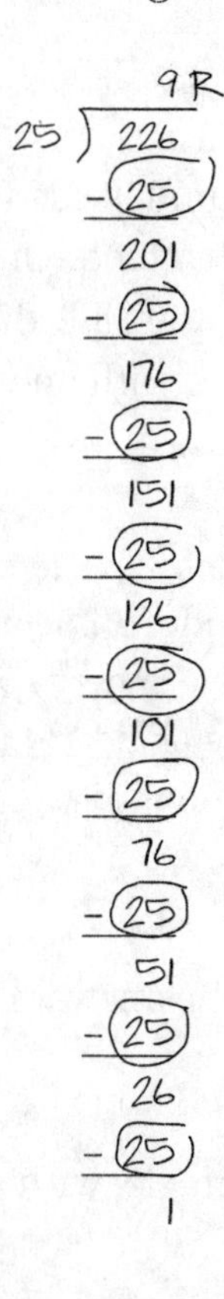

You can see in this example that 25 was repeatedly taken away from 226 nine times with one left over, but this was *not* an efficient process. However, repeated subtraction *is* a method to divide.

Another method your child might have learned in elementary school is called partial quotients, or "chunking," as it sometimes called. This is a method that has students estimate how many times the divisor goes into the dividend, and continue repeating this process until nothing remains. Here's the same problem, shown using the partial quotient method.

```
          9 R1
25 ) 226 |
   - 100 | 4
     126 |
   - 100 | 4
      26 |
    - 25 | 1
       1 |
```

The right-hand column is used to keep track of the number of times the divisor went into the dividend. Once you have completed the process, you add all the numbers in the right-hand column, in this case $4+4+1=9$, so 25 fit into 226 nine times with one left over.

Sixth-grade students are expected to write their final answer without the use of the R for remainder. Students will either take this out to a decimal answer, or write the remainder as a fraction, using the amount left over as the numerator and the divisor for the denominator. Students should then simplify the fraction, if possible.

Instead of leaving the answer as 9 R1, students should write it as a fraction: $9\frac{1}{25}$.

Here is another example, with an even larger number.

$$130\tfrac{14}{18} = 130\tfrac{7}{9}$$

```
18)2354
  - 360   | 20
   1994
 - 1800   | 100
    194
  - 180   | 10
     14
```

Depending on how you estimate how many times the divisor goes into the dividend, this process could be lengthy.

Your head is probably spinning again, thinking, "Why make it so difficult?" For years, teachers have been feeling the same way, and now sixth-grade teachers across the country are shouting "Hooray!" for this standard. It requires students to know the traditional algorithm. So here it is.

Division is asking a basic question: How many times does one number go into another number? Using the traditional algorithm, we can answer the question. How many times does 18 go into 2,354, more formally written as $2354 \div 18 = ?$

$$0130\tfrac{14}{18} = 130\tfrac{7}{9}$$

```
18)2354
  -18↓
   055
   -54↓
     14
    - 0
     14
```

This method probably looks at lot more familiar. It is a more efficient approach to dividing whole numbers. If your memory fails you and you forget how to do this, we are going to break it down into individual steps to follow in a new example. Basically, if you can remember Divide, Multiply, Subtract, Bring Down, Repeat, you will be in good shape.

$$6{,}504 \div 12$$

1. **Step 1:** Set up the division problem.

```
12)6,504
```

2. **Step 2:** How many times does 12 go into 6? Zero times.

```
      0
12 ) 6,504
```

3. **Step 3:** How many times does 12 go into 65? 5 times.

```
      05
12 ) 6,504
    -60↓↓
      504
```

4. **Step 4:** How many times does 12 go into 50? (This is the remainder after completing the subtraction.)

```
      054
12 ) 6,504
    -60↓↓
      504
     -48↓
       24
```

5. **Step 5:** Continue to repeat Divide, Multiply, Subtract, Bring Down until the division problem is complete.

```
      0542
12 ) 6,504
    -60↓↓
      504
     -48↓
       24
      -24
        0
```

The answer to 6,504 ÷ 12 is 542.

Success! Review this skill as often as your child will let you, as it doesn't go away; in fact, it becomes a lot more difficult. In algebra, your child will be using long division to factor polynomials. If he understands the process now, it will be much easier then.

Decimal Operations

The most fundamental reason to know and understand how to work with decimals is simply because of money! Our lives depend on being able to work with money to calculate total costs, the amount of change owed to us, how much it costs for multiple items of the same thing, or the cost per item when we buy in bulk. Students should understand decimals to be functional in their adult life, if for no other reason.

Adding and Subtracting Decimals

Understanding place value plays a substantial role in adding and subtracting decimals. Numbers with the same place value need to be aligned to be able to add or subtract, because they carry the same weight. But what if the decimal numbers have a different number of digits?

Students get so used to aligning whole numbers to the right that they think the same applies to decimals. However, this is not true. Corresponding place values need to be stacked one on top of another.

The rules for adding and subtracting decimals are exactly the same. The key to success is to make sure that the decimal points are lined up every time. Once the decimal points are lined up, add or subtract as you would with whole numbers. To answer the question about numbers with a different number of digits: annex zeros. Annexing zeros is a fancy way of saying it is okay to add zeros onto the end of the number after the decimal point. Adding zeros at the end of a number, after the decimal point, will not change its value. Let's see it in practice.

$$23.15 + 2.8 = ?$$

23.15
+ 2.80 ← Notice a zero was added on so that each place value lined up.
25.95 ← Bring the decimal point straight down into the sum.

Now, let's try a subtraction problem.

$6 - 4.78 = ?$

$$
\begin{array}{rcr}
6. & & \overset{5}{\not{6}}.\overset{9}{\not{0}}\,{}^{1}0 \\
\underline{-\ 4.78} & \rightarrow & \underline{-\ 4.78} \\
 & & 1.22
\end{array}
$$

Another place students make mistakes is with regrouping. They either forget to carry the one to the left when adding numbers, or forget how to regroup and carry to the right when borrowing.

Addition and subtraction of decimal problems themselves are quite simple once you know the rules. Completing the calculations requires a low level of cognitive demand, meaning it is a basic problem. To amp up the level of difficulty and the amount of thought required, students may be asked to find the error in a problem. Not only does this require procedural knowledge, but also conceptual knowledge, as an explanation is often required.

Find the error: Jane and John are finding $6.8 - 3.65$. Jane thinks the answer is 3.15 and John thinks it is 3.25. Who is correct? Explain your reasoning.

Jane is correct. The answer to this problem is 3.15. John forgot to borrow from the tenths place, so that he could subtract five hundredths from zero hundredths, a common mistake. He could have checked his work by adding 3.25 and 3.65, and would have found the sum to be 6.9, which is not what he had to start with, letting him know he made a mistake.

Multiplying Decimals

Up until this point, the majority of the experiences that students have had while multiplying have been with whole numbers. Students learn to associate multiplying with making numbers larger, because that is what has happened; for example, $2 \times 3 = 6$, $9 \times 5 = 45$, $7 \times 11 = 77$. Now we introduce decimals into the game, and things change.

$$0.5 \times 20 = 10$$
$$0.25 \times 12 = 3$$

What has happened? The product in the two examples is smaller than the second factor. Why?

In order for students to be successful with multiplying decimals, they must have a conceptual understanding of the process. A decimal less than 1 represents part of a whole. If you multiply something less than 1 by a whole number greater than or equal to 1, you are only going to get a part of it. If you are multiplying something less than 1 by another number less than 1, the same rule still applies. Let's look at a model to support this.

Model 0.7×0.6 using decimal models.

We will first begin by creating a 10×10 grid, representing 1.00, where each little square represents 0.01.

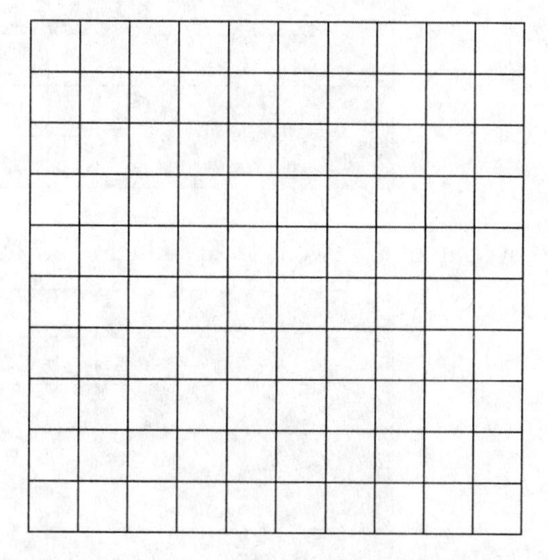

Next, we shade in 7 rows to represent 0.7 (or the first factor of the multiplication problem).

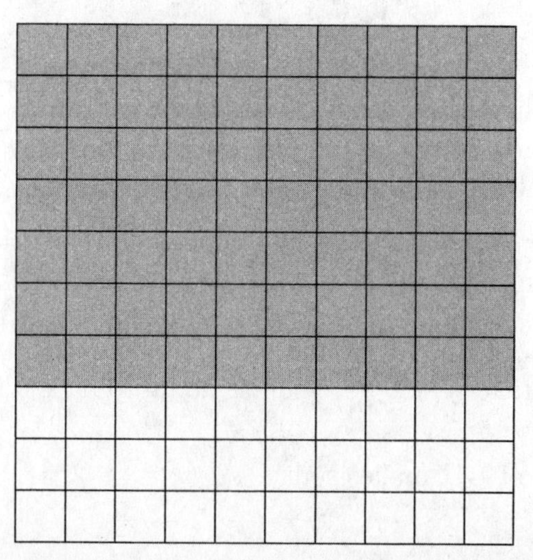

Now, we shade in 6 columns to represent 0.6 (or the second factor of the multiplication problem).

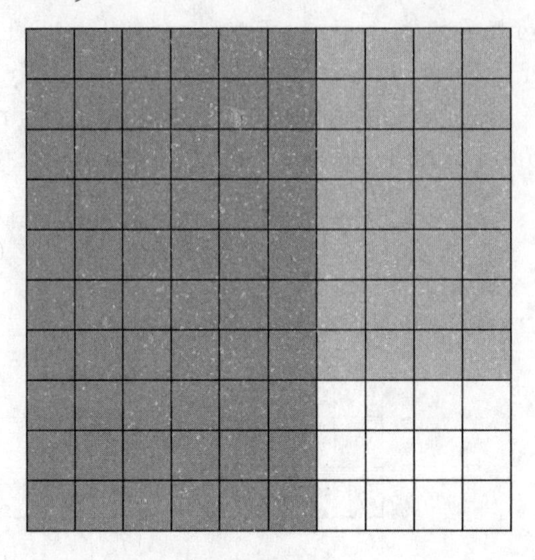

The overlap of the two is the product of 0.7 and 0.6.

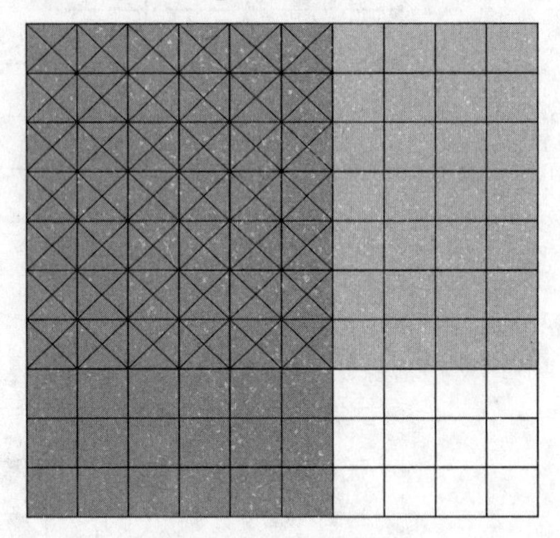

There are forty-two hundredths in the region that is marked with X. This means that the product of 0.7 and 0.6 is 0.42.

Using mathematical reasoning, does this answer make sense? Can a part of a part be greater than either of the numbers? No. This problem is saying take $\frac{7}{10}$ of $\frac{6}{10}$; you are going to have a smaller amount, so yes, the answer of $\frac{42}{100}$ makes sense.

What about if we have a decimal less than 1 times a number greater than 1?

Model 0.3×2.5 using decimal models.

We first begin by creating three 10×10 grids and shade three rows to represent the 0.3.

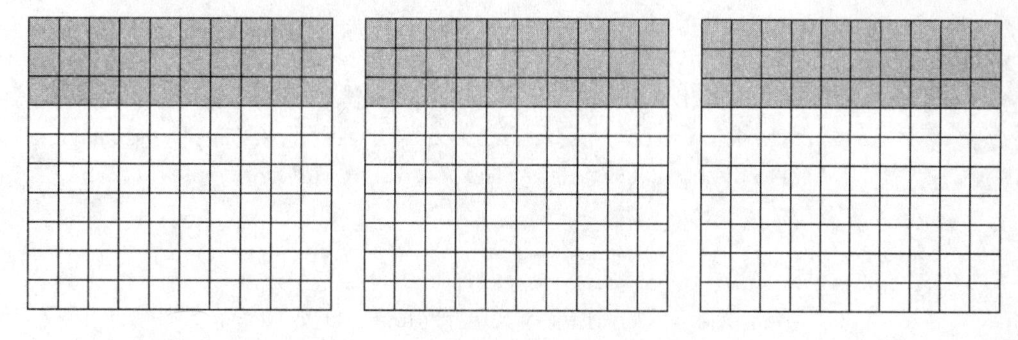

Next, we shade in all of the 2 large squares, and 0.5 of the columns in the third grid to model 2.5.

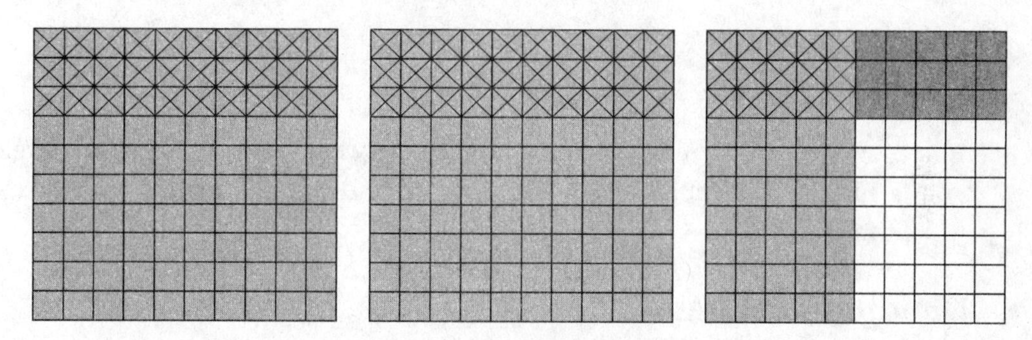

The overlap of the two represents the product of 0.3 and 2.5. Count the number of squares representing 0.01, or re-arrange the squares to see it in one square. This represents your product.

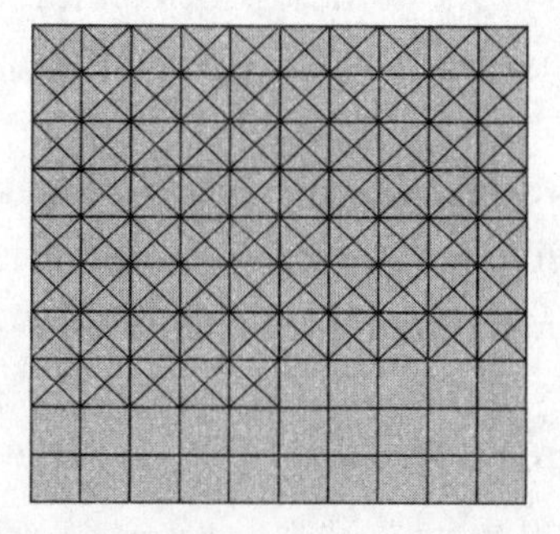

75 squares are shaded, representing a product of 0.75.

If models are too much, the rule for multiplying decimals is to multiply as you would with whole numbers. To determine where you place the decimal point in the product, simply count the number of decimal places in each factor. The total number of decimal places in the factors is the amount of decimal places needed in your product.

When multiplying decimals, you do not need to line up the decimal points. In fact, doing so only creates more computation work. Simply align all numbers to the right, and multiply like you would with two whole numbers.

Determine the product of 5.04 and 3.6.

$$\begin{array}{r} 5.04 \\ \times\ 3.6 \\ \hline 3024 \\ +\ 15120 \\ \hline 18.144 \end{array}$$

5.04 ← Two decimal places
× 3.6 ← One decimal place
18.144 ← Three decimal places needed in the product

Again, learning the procedure is pretty easy. The challenge is developing the conceptual understanding to retain the information for long-term mastery.

Dividing Decimals

Dividing decimals is a place where many students hit a roadblock. One piece of advice to help students be successful while practicing this skill is to use graph paper to help align the numbers when dividing. If students are able to keep their work in nice, neat columns, it will help to keep all the digits and the decimal point in the appropriate place value.

Dividing Decimals by Whole Numbers

Dividing decimals by whole numbers is very similar to dividing two whole numbers. We understand breaking an amount into two parts (dividing by 2) or three parts (dividing by 3), for example, so $3.5 \div 7 = 0.5$ makes sense. You are breaking 3.5 into 7 equal-sized groups. Each group must be

0.5 in size. To check that you did your calculations correctly, you can use the inverse operation to division, which is multiplication, to check your work.

Here is the rule: To divide a decimal by a whole number, place the decimal point directly above the decimal point in the dividend, and divide as you would with whole numbers. Pretty easy, right?

Dividing Decimals by Decimals

Here is where it gets tricky. The Common Core Standards expect students to fluently divide decimals using the standard algorithm. We already know the standard algorithm for dividing two whole numbers. We have applied this knowledge to dividing decimals by whole numbers; now we need to take it one step further. The good news is that the algorithm doesn't change. The bad news is the work leading up to it does.

Our brains don't tend to think of things in terms of fractional parts of something, so dividing any number by 0.12 is challenging, but can you divide a number by 12? Absolutely. This is how we approach dividing by a decimal number. How much larger is 12 when compared to 0.12? If you said 100 times larger, you would be correct.

When dividing by decimals, we change the divisor into a whole number by multiplying by a power of 10. In order to maintain the same ratio between the divisor and the dividend, we must multiply the dividend by the same power of 10.

$$1.86 \div 0.12$$

By multiplying both the dividend and the divisor by 100, the problem becomes $186 \div 12$, a much easier version to think about.

Let's take the next problem all the way through.

$$14.19 \div 2.2$$

What power of 10 do we need to multiply 2.2 by to make it a whole number? If you said 10, you'd be correct. Remember, whatever we do to the divisor, we must also do to the dividend.

$$2.2 \bullet 10 = 22$$
$$14.19 \bullet 10 = 141.9$$

And the problem becomes $141.9 \div 22$. Let's set this up.

The first number is referred to as the dividend, and this goes under the division symbol, like so:

$\overline{)141.9}$

The second number is the divisor and this goes on the outside.

$22\overline{)141.9}$

Now, bring the decimal point directly up into the quotient (the answer to a division problem) and divide as you would with whole numbers.

```
      6.45
22)141.90
  -132↓ |
     99 |
    -88 ↓
     110
    -110
       0
```

Again, once you have mastered how to divide whole numbers, this part is not so bad. Encourage your child to practice this skill as often as possible, even if he is not using it in class. Remember, practice makes progress.

Greatest Common Factor

Students will learn to find the greatest common factor of two whole numbers less than or equal to 100. This skill will help students simplify fractions now, and will help them in later grades when they are learning to factor polynomials.

To begin, let's think about what the words mean. Some synonyms for *greatest* are largest, biggest, or maximum. *Common* means shared. *Factor* is part of a multiplication problem. So in determining the greatest common factor of two numbers, we are looking for the largest factor shared by two (or more) numbers.

There are two methods we can use to find the greatest common factor (GCF) of two or more numbers: the listing method, and prime factorization with Venn diagrams.

Method 1: The Listing Method

In the listing method, you begin by writing all of the factors for each number. Let's try this for 24 and 18.

24: 1, 2, 3, 4, 6, 8, 12, 24
18: 1, 2, 3, 6, 9, 18

The factors that 24 and 18 share in common are 1, 2, 3, and 6, but the greatest common factor is 6 because it is the largest factor shared by both.

Method 2: Prime Factorization and Venn Diagrams

Using factor trees to determine the prime factorization of each number is the first step in this method.

A prime number is any number that has exactly two factors, 1 and itself.

Let's use the same numbers as before, 24 and 18.

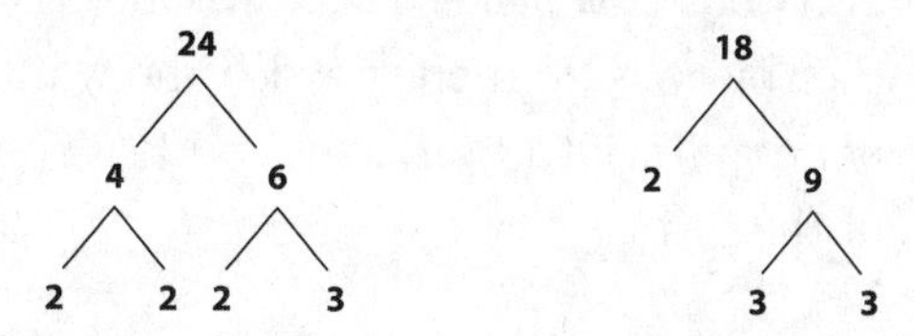

The prime factorization of 24 is $2 \bullet 2 \bullet 2 \bullet 3$ or $2^3 \bullet 3$

The prime factorization of 18 is $2 \bullet 3 \bullet 3$, or $2 \bullet 3^2$

Knowing the prime factorization of these two numbers allows us to build a Venn diagram.

$24 = \mathbf{2} \bullet 2 \bullet 2 \bullet \mathbf{3}$

$18 = \mathbf{2} \bullet \mathbf{3} \bullet 3$

24
18
2
2
2
3
3

The factors are placed into each part of the Venn diagram. The factors shared in common are placed where the two circles overlap. You can see that a factor of 2 and a factor of 3 are both shared in common. By multiplying the common factors together, you are able to determine that the GCF of 24 and 18 is 6.

Once again, asking students to find the GCF of two numbers like the previous problems are procedural-level questions not requiring much thought. In an effort to increase students' level of understanding and make the problems more challenging, word problems are introduced.

A student can tell if it is a GCF word problem by looking for these key elements:

1. We have to split things into smaller sections.
2. We are trying to arrange something into rows or groups.
3. The question asks to find the "greatest," "biggest," or "largest" number of something.

Here are some sample questions you might come across that require solving for the GCF.

> Samantha has two pieces of cloth. One piece is 72 inches wide, and the other piece is 90 inches wide. She wants to cut both pieces into strips of equal width that are as wide as possible. How wide should she cut the strips?
>
> Bennie is catering a party and putting snack food on plates. He has 72 cheese puffs and 48 carrot sticks. He wants both kinds of food on each plate. He wants to distribute the food evenly and he doesn't want any left over. What is the largest number of plates he can use, and how many of each type of food should he put on each plate? Show your work and explain your answer using vocabulary from our study of number theory.

Notice in these types of problems that there is a greater demand of the student to think about what needs to be done to find an answer. Also, an explanation is required, which develops a greater level of understanding for the topic. Questions such as these were developed and shared by Deb LaChance, a sixth-grade math teacher at Oyster River Middle School.

Least Common Multiple

Just like we did for GCF, let's do the same for Least Common Multiple and break down what the words mean. Synonyms for *least* include tiniest, smallest, or minimum. We already know what the word *common* means, but now we have figure out what *multiple* means. In this case, we are talking about the multiples of a number. The multiples of 2 are 2, 4, 6, 8, 10, 12. The multiples of 3 are 3, 6, 9, 12, 15, 18. When you hear the word multiple, think of a multiplication grid.

So, the least common multiple (LCM) means the smallest multiple shared by two numbers. In the case of 2 and 3, the LCM is 6. It is the first time that they have a multiple in common.

Just as we had two methods for GCF, there are two methods for LCM, and although they share the same titles, they do look a bit different. Let's use the two methods to find the LCM of 12 and 18.

Method 1: The Listing Method

Instead of listing the factors of two numbers like you did for GCF, this listing method requires you to list the multiples of the two numbers.

The multiples of 12 are: 12, 24, 36, 48, 60, 72, 84, 96, 108 . . .

The multiples of 18 are: 18, 36, 54, 72, 90, 108 . . .

12 and 18 share the following multiples in common: 36, 72, 108 . . . . The least common multiple is 36, because it is the smallest number in the list.

Method 2: Prime Factorization and Venn Diagrams

This method starts out the exact way as it does for GCF. You begin by finding the prime factors of the two numbers using factor trees. Then you place them in a Venn diagram, putting shared factors in the middle and the others in each of the respective circles.

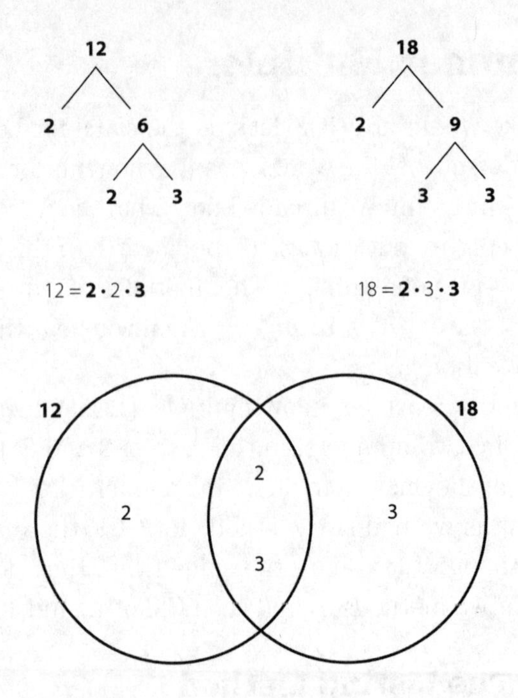

The GCF of these two numbers is 6, found by multiplying the numbers in the center together. The LCM is found by multiplying *all* of the factors in the Venn diagram. In this case the LCM(12, 18) = 2 • 2 • 3 • 3 = 36.

Least Common Multiple Understanding

Is one solving method better than the other? No, it comes down to personal preference. If your child were given a problem that asked her to find the GCF and the LCM of the same two numbers, a suggestion would be to use the second method because you have done the work for both in creating the Venn diagram. However, if the listing method makes sense, you child can list factors in one list to get the GCF, and the multiples in another list to get the LCM.

The basic LCM questions do not really meet the criteria as set by the Common Core State Standards. The goal is to develop students with strong mathematical understanding and problem-solving skills. To reach this goal, we once again turn to word problems.

It is a LCM problem if:

1. We have an event that is or will be repeating over and over.
2. We have to get multiple items in order to have enough.

3. We are trying to figure out when something will happen again at the same time.
4. The final question asks to the find the "smallest" or "least" number of something.

Here are some sample questions you might come across that require solving for the LCM.

> Ben exercises every 12 days, and Isabel every 8 days. Ben and Isabel both exercised today. How many days will it be until they exercise together again?
>
> A local restaurant is offering a free meal to every 25th customer and a free hat to every 12th customer. Which customer will be the first to get a free meal *and* a free hat? Show your work and explain your answer using vocabulary from our study of number theory.
>
> Find the LCM of the factors of 200. Show your work and explain your answer.

Don't be afraid to watch your child struggle a bit when solving these problems. She needs a chance to think about what is being asked, and then consider what approach is best. A little struggle will lead to greater understanding in the end.

Integers

The final piece of the Number System in grade 6 is to introduce students to signed numbers called integers. Students will learn how integers are used to represent situations, how they can be ordered, and how they are used to graph on the coordinate plane.

Integers are best defined as any whole number and its opposite. *Integers* are the set of positive and negative whole numbers, including zero (although this has no opposite). *Positive integers* are to the right of 0 on a number line. *Negative integers* are to the left of 0 on a number line. Opposite integers are the same distance from 0, on opposite sides. For example, 5 and -5 are opposites, because they both have a distance of 5 from zero on the number line.

As sixth graders, students need to learn the role of an integer. They can be used to describe quantities that can have both positive and negative values. Here is a list of some real-life examples where integers are commonly used:

- Temperature (above and below 0)
- Elevation (above and below sea level)
- Money (credits and debits, profits and losses)
- Electric charges (protons, electrons)
- Football field (gain and loss of yards)

How could you use an integer to describe the following?

1. A temperature of 15 degrees below (Fahrenheit) –15 degrees F
(The word "below" indicated a negative amount.)
2. A gain of 8 yards 8 yards
3. A debt of $42 –$42

When an integer is positive, it is not necessary to put a + symbol in front of the number.

Ordering Integers

Ordering whole numbers greater than zero is an easy task: The bigger the number, the farther to the right it lies. By this point, students have been working with whole numbers for quite some time and generally have a solid understanding of the magnitude of a whole number. It makes sense that when ordering numbers from least to greatest, those closest to 0 come first, and those farthest away from 0, go last.

Now we bring in negative numbers, and it is not so easy. To order integers, we need to look at the magnitude of a number. Another way to measure magnitude is to determine its absolute value, or its distance from zero on a number line. The symbol used to find absolute value is a pair of vertical parallel lines, $|n|$. Since absolute value measures distance, the result is always positive.

$|-4| = 4$	-4 is a distance of 4 from 0 on the number line
$|8| = 8$	8 is a distance of 8 from 0

The greater the absolute value of a negative number is, the farther to the left on the number line it lies, meaning it is the least number.

Let's order the following set of numbers from least to greatest:

$$\{-18, 36, 14, 0, -25, 21, -7\}$$

-25 has a greater absolute value than -18, which means -25 carries less value than -18. So the first number in the list is -25. Using this same level of comparison for each pair of numbers in the set, you can determine the ordering.

The numbers in order from least to greatest are:

$$-25, -18, -7, 0, 14, 21, 36$$

The farther to the right a number lies on a number line, the greater value it has.

Plotting Integers on a Number Line

Integers are also used to represent a location, either on a number line or on a coordinate plane. Let's consider the integer -3. It is located three places to the left of 0. Its opposite is located three places to the right, and has a value of 3. What is the opposite of 12? If you said -12 you are correct. The number 12 is twelve places to the right of 0, and its opposite -12, is located twelve places to the left of 0. What would the opposite of the opposite of a number be? This is expressed as $-(-a)$. Let's try $-(-7)$. If the opposite of 7 is -7, then the opposite of the opposite would be the number itself, 7. Boy, that can be confusing, but it really just comes down to the fact that a double negative makes a positive.

The Coordinate Plane

The *coordinate plane* is formed by the intersection of a horizontal number line (called the x-axis) and a vertical number line (called the y-axis). The point of intersection is called the *origin*, and is defined by the ordered pair (0, 0).

When these two number lines intersect, they break the coordinate plane into four quadrants. Refer to the diagram for the names of each quadrant.

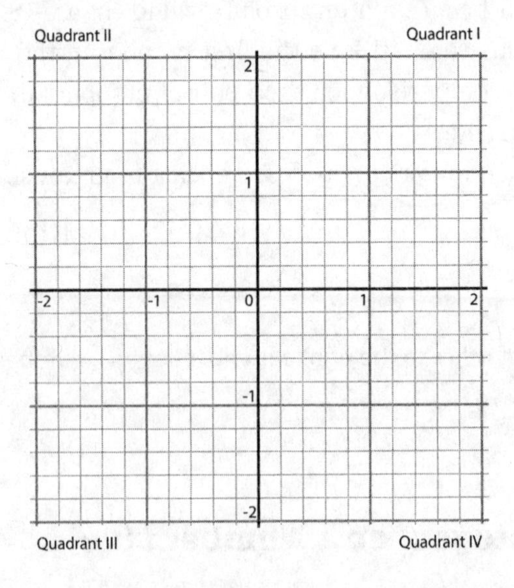

A common mistake is that students name the quadrants in a clockwise rotation, rather than the correct counterclockwise rotation.

You can use an ordered pair to name any point on the coordinate plane. To mark a location on a coordinate plane, integers are used to create ordered pairs: (x, y). The first number, known as the x-coordinate, tells the location on the horizontal number line; the second number, known as the y-coordinate, tells the location on the vertical number line. The intersection of these two points marks the location of the point on the coordinate plane. The ordered pair (5, –3) informs the student to move 5 units to the right and 3 units down. This places this point in Quadrant IV.

All points in Quadrant I have both a positive x-coordinate and positive y-coordinate. The points in Quadrant II have a negative x-coordinate and a positive y-coordinate. The points in Quadrant III have a negative x-coordinate and a negative y-coordinate. Lastly, the points in Quadrant IV have a positive x-coordinate and a negative y-coordinate.

Identify the quadrant where the point is located.
(–3, 2) is in Quadrant II
(–4, –6) is in Quadrant III
(4, 7) is in Quadrant I
(8, –2) is in Quadrant IV

If the x-axis or the y-axis forms a line of symmetry between two points, the two points are considered to be reflections of one another. If the x-coordinate remains the same, and the y-coordinates are opposite each other, the two points have been reflected over the x-axis. If the y-coordinate remains the same, and the x-coordinates are opposite each other, the two points have been reflected across the y-axis. The distance between the two points can be found by adding the absolute value of the coordinate with opposite integers. For example, if you are given the ordered pairs (3, 5) and (3, –5), the distance between the two points is found by the sum of $|5|+|-5|=5+5=10$ units. These two points have been reflected over the x-axis and are 10 units apart.

How have these two ordered pairs been reflected and what is the distance between them? (4, –6) and (–4, –6)

These points were reflected over the y-axis and are 8 units away from each other.

Practice Problems

Determine the quotient of each of the problems.

1. $\frac{2}{3} \div \frac{1}{2}$

2. $8 \div \frac{3}{4}$

3. $8\frac{3}{4} \div 2\frac{1}{3}$
4. $\frac{3}{8} \div \frac{1}{6}$
5. $7{,}800 \div 24$
6. $390 \div 26$
7. $182 \div 13$
8. $18{,}408 \div 52$

Determine the sum or difference.

9. $62.38 + 18.96$
10. $81.4 + 13.92 + 2.678$
11. $58.9 - 13.74$
12. $1{,}324 - 562.32$

Determine the product.

13. 0.45×3.8
14. 2.75×0.9
15. $\$3.56 \times 5$

Determine the quotient.

16. $3.78 \div 7$
17. $1.725 \div 0.05$
18. $0.875 \div 0.2$

Determine the GCF or LCM for each question.

19. GCF(12, 32)
20. LCM(2, 8)
21. If the LCM of 12 and some number is equal to 36, find three possible values for the missing number.
22. Anna sells bags of different kinds of cookies. She made $27 selling bags of peanut butter cookies, $18 selling bags of chocolate chip cookies, and $45 selling bags of oatmeal cookies. Each bag of cookies costs the same amount. What is the most that Anna could charge for each bag of cookies?

Write an integer to represent each of the situations.

23. A debt of $30
24. An increase in temperature of 18 degrees
25. 25 meters below sea level
26. A gain of 25 yards

Determine the absolute value.

27. $|-15|$
28. $|18|$
29. $|-6|$
30. $|-5|+|7|+|-12|$

Write an ordered pair to represent the location described.

31. Move 3 units left, and 5 units down.
32. Move 6 units right, and 4 units down.
33. Move 1 unit right, and up 4 units.
34. What is the location of the point (2, −4) after it has been reflected over the x-axis?
35. What is location of the point (4, 5) after it has been reflected over the y-axis?
36. What is the distance between (2, 8) and the reflection of this point over the x-axis?

CHAPTER 7

Expressions and Equations

It is in the sixth grade that students understand the use of variables in expressions and equations, and use this understanding to solve problems. Students also learn the formal way in which to solve an equation, treating it like a balanced scale. Establishing a solid understanding of expressions and equations in sixth grade will allow your child to be successful in more advanced math classes. Honestly, there are so many details that could be included in this chapter that it could take up the whole book. This chapter will attempt to keep it brief, but informative.

What Should My Child Already Be Able to Do?

In fifth grade, students learn to write and interpret numerical expressions, including the use of parentheses, brackets, and braces. They are able to take basic word statements and transform them into numerical expressions: for example, "3 more than 2" becomes the expression $2+3$. If a student were given a set of directions, such as "add 4 and 6, then multiply by 5," it is expected that the student write the expression $5\times(4+6)$. The goal of the Common Core State Standards is to have students view $5\times(4+6)$ as five times as large as the sum of $4+6$, without having completed any computations at all.

Students in fifth grade are also expected to analyze patterns and relationships based on two given rules. For instance, if a child were given the rule "add 2," he should know that for every number he is given, he should add 2 to find the sum. This is the foundation for learning about input/output tables, and eventually functions.

What Does an Exponent Do?

Much like multiplication is a more efficient means for repeated addition, *exponents* are a more efficient way to represent situations with repeated multiplication. This type of number is called a *power*.

$$2\times2\times2\times2=2^4$$

Here the exponent is 4 and is used because the 2, also known as the base, is being multiplied by itself four times. 2^4 is called a *power*.

So how do you figure out what exponent to use? First, identify what number is being multiplied repeatedly. This is your *base*. The number of times your base is used as a factor tells you what number to use as the exponent.

The word base has two meanings in mathematics, depending on the context in which it is being used. With powers and exponents, the base is the factor being multiplied repeatedly. In geometry, the base is one of two parallel sides (2-dimensional shape) or faces (3-dimensional shape).

Try this problem:

$$3 \times 3 = ?$$

The base is ________? (Hopefully, you said the base is 3.)
It is being used as a factor ____ times. (It is used as a factor 2 times.)
So the exponent is ____. (Exponent is 2).
And the power is written as ____. (Written as a power: 3^2)

Here is some other important information to know. Any number raised to the power of 1 is the number itself. $5^1 = 5$. For this reason, the power of 1 is not used, as it does not change the value of the expression. Interestingly enough, one might think that any number raised to the zero power is 0; however this is *not* true. In fact, any number raised to the power of 0 is 1. Using a pattern, we are able to see this more clearly.

$2^4 = 16$
$2^3 = 8$ $(16 \div 2 = 8)$
$2^2 = 4$ $(8 \div 2 = 4)$
$2^1 = 2$ $(4 \div 2 = 2)$
$2^0 = 1$ $(2 \div 2 = 1)$

By using the pattern, you can see that with each decrease in power the base divides the previous product to get the value. When you divide 2 by 2, you get 1, proving that $2^0 = 1$. Cool, huh? This pattern holds true no matter what number is used for the base.

Commoncore.org states that your child should be "able to write and evaluate numerical expressions involving whole-number exponents." What does this look like?

Writing the expressions is often the easy part, because students can count the number of times the base is used as a factor and write the expression.

$2 \times 2 \times 2 \times 2 \times 2 \times 2 = 2^6$
$9 \times 9 \times 9 \times 9 \times 9 = 9^5$
$10 \times 10 \times 10 \times 10 \times 10 \times 10 \times 10 \times 10 = 10^8$
$2 \times 2 \times 2 + 4 \times 4 \times 4 \times 4 = 2^3 + 4^4$

When the directions change, it can become a bit more difficult. What if you're asked to write each power as a product of the same factor, then find the value? When writing a power as a product of the same factor, that means you are writing it in expanded form, or as a multiplication problem showing how many times the base is used as a factor. Find the value means determine the answer.

$2^3 = 2 \times 2 \times 2$ ← the base (2) was used as a factor 3 times
4×2
8 ← This is the value

A common mistake made when students are calculating the value is to try a shortcut and not write it as a product of the same factor. This leads to students turning 2^3 into $2 \times 3 = 6$, rather than the correct way of $2 \times 2 \times 2 = 8$.

To help students remember what an exponent does, you can use this silly phrase: "An exponent tells the base how many times to show its face."

Powers and exponents are used in a variety of contexts, ranging from expressions that involve powers and require the use of order of operations, to geometry concepts involving two-dimensional and three-dimensional shapes, all the way to algebraic concepts involving polynomials and the quadratic equation. Knowing what they mean and how to interpret them is of utmost importance.

Writing and Evaluating Expressions

Students move from very concrete expressions to more abstract algebraic expressions where variables are used, very quickly in grade 6. The basic expression "three more than two," or $2+3$, leads to the algebraic expression $n+3$, when the verbal expression "three more than a number" is used. Learning what words and phrases indicate which operations is the first element to understanding how to write and evaluate expressions. Here is a table that outlines the most commonly used words, and to what operation they apply.

▼ WORDS AND OPERATIONS

Add	Subtract	Multiply	Divide
Sum	Difference	Product	Quotient
Plus	Minus	Times	Divide
Total	Less	Twice	Per
More than	Less than	Triple	Each
Increase	Decrease	Double	Half
In all	Remains	Area	Ratio
Raised	Lowered	Multiplied	Broken into
Altogether	Reduce	Every	One
Older than	Younger than	Square	One-fourth

This list could go on and on, but you get the idea. If you become familiar with these words, then you are able to associate one or more of their synonyms to the operation.

Now that you know what words indicate particular operations, you need to learn how to write an algebraic expression. An *algebraic expression* is a combination of variables, numbers, and at least one operation.

A variable is a symbol, usually a letter, used to represent an unknown number. In younger grades a square, circle, or triangle would be placed into an equation to represent an unknown value. In an effort to prepare students for algebra, middle-school students use x, y, a, b, c, or any other letter of the alphabet to represent the unknown number.

The idea of being able to write an algebraic expression from a verbal/written statement is helpful when solving word problems and real-world scenarios. There are three steps to follow when helping students learn this skill:

1. **Step 1:** Define the variable. What is the unknown representing?
2. **Step 2:** Identify the operation being used. What are the keywords telling you?
3. **Step 3:** Write the algebraic expression based on the given information.

Johnny's age is 5 years less than his brother.

1. **Step 1:** Let b = age of Johnny's brother
2. **Step 2:** "Less than" indicates it is a subtraction problem
3. **Step 3:** The algebraic expression representing Johnny's age is $b-5$

The cost of milk at a convenience store is 3 times as much as at a grocery store.

1. **Step 1:** Let c = cost of milk at grocery store
2. **Step 2:** "Three times as much" means to multiply by 3
3. **Step 3:** The algebraic expression representing the cost of milk at the convenience store is $3c$

There are two words for subtraction that are very similar in name, but have very different meanings. These two phrases are *less than*, and *less*. "7 less than a number" is the algebraic expression $x-7$. "7 less a number" is $7-n$. Do you notice the difference? This can be very confusing for students. Remember to read the phrase, put a value in for the variable, and evaluate the expression. If the answer makes sense, then it is probably correct.

Sometimes it is necessary to use parentheses to ensure that the intent of the calculations in the verbal expression are met when evaluating the algebraic expression using the order of operations. For example, if the verbal expression were "four times the sum of 5 and a number," one might write $4\times5+n$, while another might write $4\times(5+n)$. Which expression is correct? Let's set $n=2$. The order of operations says to do all multiplication and division before adding or subtracting. The first expression would simplify to $20+2$, or 22. The second expression would simplify to 4×7, or 28. What the verbal statement is saying is "four times the sum of 5 and 2." This verbal statement requires you to add first and then multiply, requiring the use of parentheses. So, $4\times(5+n)$ is the correct algebraic expression.

Evaluating expressions is often a favorite of sixth-grade students. Once given the structure of how to organize and show their work, they like the ability to take a complex expression and evaluate it, simplifying it to a single value. Using these steps, students find great success.

1. **Step 1:** Copy original problem.
2. **Step 2:** Substitute the given value(s) of the variable(s).
3. **Step 3:** Simplify using order of operations (work should be in the shape of an ice cream cone).

Here is an example:

Evaluate $z^2 + 5y - 20$, where $y = 15$, and $z = 8$

1. **Step 1:** Copy the original problem (especially if it is in a textbook).

$$z^2 + 5y - 20$$

2. **Step 2:** Substitute the values for the variables. $8^2 + 5(15) - 20$
3. **Step 3:** Simplify using order of operations.

Notice how it looks like an ice cream cone.

$$64 + 5(15) - 20$$
$$64 + 75 - 20$$
$$139 - 20$$
$$119$$

Knowing how to substitute and simplify expressions will be important once students start working with formulas such as distance = rate × time, area of a rectangle = base × height, volume of a rectangular prism = length × width × height, etc. All of these formulas require values to be substituted for the variables, and then expressions can be simplified.

Properties of Operations

Mathematical properties of operations are a set of rules that allow someone to change the look of an expression, but not change its value; they are called equivalent expressions. When work begins with algebraic expressions, it is often necessary to apply these properties to be able to simplify the expression.

▼ PROPERTIES OF OPERATIONS

Name of Property	Definition	Example
Commutative Property of Addition	The order in which numbers are added does not change the sum.	$2+3=3+2$ $5=5$
Associative Property of Addition	The way in which numbers are grouped when added does not change the sum.	$(3+4)+5=3+(4+5)$ $7+5=3+9$ $12=12$
Identity Property of Addition or Additive Identity	The sum of any number and 0 is the original number.	$6+0=6$
Commutative Property of Multiplication	The order in which numbers are multiplied does not change the product.	$4 \times 5=5 \times 4$ $20=20$
Associative Property of Multiplication	Changing the grouping ( ) of the numbers will not change the product.	$2\bullet(3\bullet4)=(2\bullet3)\bullet4$ $2\bullet12=6\bullet4$ $24=24$
Identity Property of Multiplication or Multiplicative Identity	The product of any number and 1 is the original number.	$8\bullet1=8$
Multiplicative Property of Zero	The product of any number and zero will always be 0.	$2\bullet0=0$
Distributive Property	The product of a number and a sum or difference is the same as the sum or difference of each product taken separately.	$3(7-4)=3(7)-3(4)$ $3(3)=21-12$ $9=9$ Algebraically: $a(b+c)=ab+ac$ $a(b-c)=ab-ac$

Simplifying Expressions

Mathematicians like to be able to provide a reason for why things work. The properties of operations are oftentimes their reason, or their proof. Let's simplify a few expressions and prove each step along the way.

$3+(5+x)$ 5 and x are considered *unlike terms*. They cannot be combined.

$(3+5)+x$	The *associative property* of addition allows you to change the way the numbers being added are grouped.
$8+x$	3 and 5 can be combined because they are *like terms*. The expression simplifies to $8+x$.
$4x+7+2x+3$	
$4x+2x+7+3$	The *commutative property* of addition allows you to change the order of the numbers being added. The 7 and $2x$ changed places.
$6x+10$	$4x$ and $2x$ are like terms because they contain the same variable, and when added together become $6x$. 7 and 3 are like terms, and when added together make 10.

Both of these examples begin and end with very different-looking expressions; nonetheless, they are equivalent.

The *distributive property* is a bit unique in that it combines multiplication with addition or subtraction. Think about the title of this property. What does the word distribute mean? Distribute means to deliver or pass out. Each Halloween you distribute candy to the trick-or-treaters: You hand out candy to each person who comes to your door. The mathematical property works the same way. The verbal expression "four times the sum of a number and 3" looks a bit complex when written algebraically as $4(x+3)$. By following the rules of the distributive property, you can distribute the 4 to each part of the addition problem: $4(x+3)=4\bullet x+4\bullet 3$. When this is simplified, it becomes $4x+12$. You can also use the distributive property in reverse to take a common factor out of each of the terms.

$20x+35y$	The numbers 20 and 35 share a greatest common factor of 5.
$5\bullet 4x+5\bullet 7y$	You can take the common factor out of each term.
$5(4x+7y)$	This can be rewritten as $20x+35y=5(4x+7y)$.

When Are Two Expressions Equal?

Applying properties to see if two expressions are equivalent can often be like a puzzle. What mathematically correct actions can you take to manipulate one expression to see if it looks like another? Two expressions are equivalent when the results of each expression are the same number, regardless of the value assigned to the variable.

For example, is $z+z+z+z$ equal to $4z$? Let's assign 3 to the value of z and simplify each expression to see what happens.

$z+z+z+z$	$4z$
$3+3+3+3$	$4 \bullet 3$
$6+3+3$	12
$9+3$	
12	

The results are the same, therefore $z+z+z+z=4z$. Even if the value for z changed, the results would remain equal.

Solving One-Variable Equations

Students begin to solve an equation by deciding which values from a specified set make an equation true. They do this using a familiar process of *substitution*, where they replace the variable in the equation with one of the values from the set of possible solutions. Once the equation is simplified, the value to the left side of the equal sign is compared to the value on the right. If they are equal, then the value for the variable is a solution to the equation.

What is the difference between an expression and an equation?
An *expression* is a combination of numbers and at least one operation. An *equation* is two expressions set equal to one another. So basically, equations have an equal sign, and expressions do not. Equations can be solved; expressions are evaluated.

Solving equations in sixth grade deals with a single variable on one side of the equal sign. Students learn to solve one-step equations involving a single operation (addition, subtraction, multiplication, or division) that take on the form $x+p=q$, $x-p=q$, $px=q$, or $x/p=q$. It is important for students to understand that an equation is much like a scale or balance. The two sides of the balance must remain equal at all times.

Imagine a seesaw at your local playground. The goal of the children playing on it is to get it to balance perfectly, meaning the weight on the left side of the fulcrum equals the weight on the right side of the fulcrum. Once it is balanced and a child gets off one side, what happens? The heavier side falls to the ground and the two sides are no longer equal. It is time to start over. How could we change the situation so one side didn't fall to the ground? How could we keep the seesaw balanced if we remove one person from one side? We could remove a person of the same size from the other side.

This same principle holds true when dealing with an equation. Think about the equal sign as the fulcrum. Whatever is on the left side of the equation is equal to what is on the right.

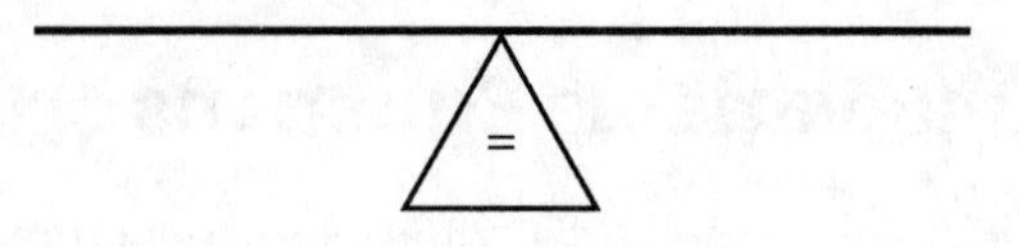

Let's solve the equation using the principal of a balanced equation.

$$x+9=14$$

By creating a model, we can get a visual representation of the equation, which will help develop conceptual understanding.

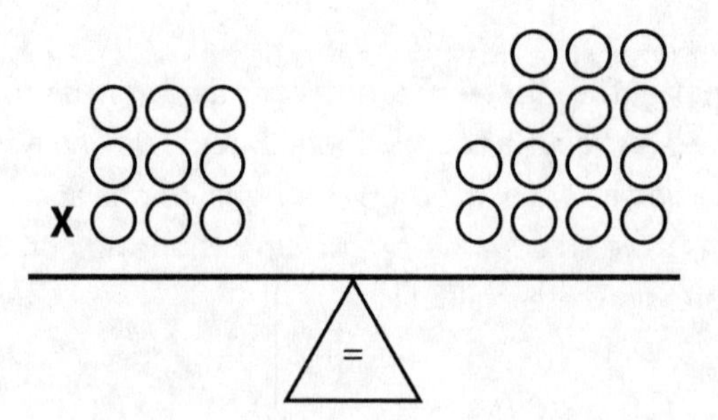

The goal of solving an equation is to figure out what x is equal to. We do this by isolating the variable, or getting the variable by itself on one side of the equation. Right now, x is on the same side as the 9. How can we get x by itself? Take the 9 away, but whatever we do to one side, we need to do to the other. Let's take 9 away from *both* sides of the equation.

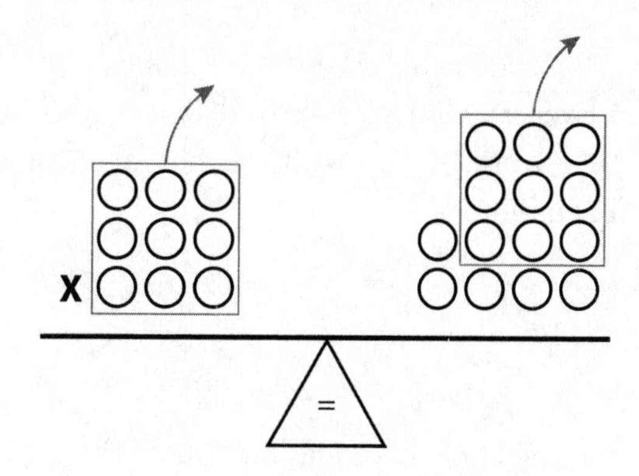

Now that we have taken 9 away from each side, what is left?

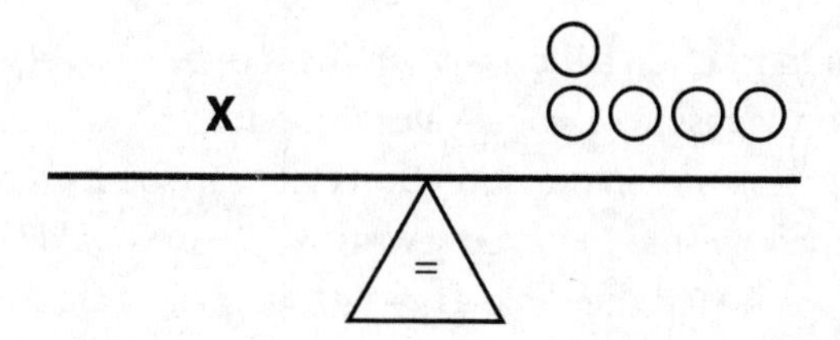

The x is left on one side of the balance, or equal sign, and 5 is left on the other side. What does this mean? This means that $x = 5$, and 5 is the solution to the equation.

Models are one way to solve equations, but the drawings can become time-consuming and tedious. Let's look at the algebraic way of solving the same equation.

$$x + 9 = 14$$

Your first step still remains the same: You need to isolate the variable. To do this, we use inverse operations, or the opposite operation. Since addition is the operation used in the equation, we will use subtraction to undo the problem. Remember, whatever you do to one side, you need to do to the other to maintain equality.

$$\begin{array}{r} x+9=14 \\ \underline{-9\;-9} \\ x=5 \end{array}$$

You solved exactly the same problem and came up with the same answer. Even though you have a solution, an important step is to always check your work by substituting the value for the variable back into the original equation.

$$\begin{array}{r} \text{Check: } x+9=14 \\ 5+9=14 \\ 14=14 \end{array}$$

The left side of the equation equals the right side! Hooray, it has been solved correctly.

Students often struggle when the format of an equation has been changed and the result, or single value, is on the left side of the equal sign and the expression is on the right. Don't let them be fooled; you treat it exactly the same way, and can even rewrite it, if you would like. For example, $21 = x - 10$ is the same as $x - 10 = 21$. This gets students every time.

Every form of an equation has a unique set of steps used to solve it. A model for each type of one-step equation has been set up for you to see.

Addition Equation:

$$\begin{array}{l} y+12=15 \\ \underline{\;\;-12\;-12} \\ \;\;y\;\;\;=3 \end{array}$$

$$\begin{array}{r} \text{Check: } y+12=15 \\ 3+12=15 \\ 15=15 \end{array}$$

Subtraction Equation:

$$\begin{array}{l} z-6=14 \\ \underline{\;\;+6\;+6} \\ \;\;z\;\;=20 \end{array}$$

$$\begin{array}{r} \text{Check: } z-6=14 \\ 20-6=14 \\ 14=14 \end{array}$$

Multiplication Equation:

$7w = 42$

To undo multiplication, division is used.

$\frac{7w}{7} = \frac{42}{7}$

$w = 6$

Check: $7w = 42$

$7(6) = 42$

$42 = 42$

Division Equation:

$\frac{m}{3} = 5$

To undo division, you multiply.

$\frac{3}{1} \bullet \frac{m}{3} = 5 \bullet 3$

$m = 15$

Check: $\frac{m}{3} = 5$

$\frac{15}{3} = 5$

$5 = 5$

As you look at these examples, you are probably thinking, "My child can solve these in her head." Although it may be true, learning how to show work in even the most basic of examples develops really strong work habits and great organization. Once the skills are established for basic problems, solving multistep equations in seventh and eighth grade is not so bad!

Inequalities

Inequalities are exactly what they say they are: a mathematical sentence indicating that two quantities or expressions are not equal. Inequalities explore expressions or quantities that are less than or greater than the other, and involve the use of the following symbols: $<$, $>$, $\leq$, or $\geq$. Inequalities can also be true or false.

Identify each of the following as true or false.

1. $4 < 7$
2. $9 > 11$
3. A's age is 56 and B's age is 49. Therefore, $A > B$
4. If $P = 20$, $Y = 15$, then $Y > P$
5. $9 \leq 9$

The answers are as follows:

1. True
2. False
3. True
4. False
5. True

Students may also be asked which inequality symbol best answers the statements. Let's try a few.

Replace each ____ with the correct symbol.

1. 5 ____ 14
2. 9 ____ 3

And the survey says:

1. 5 $\underline{< \text{ or } \leq}$ 14
2. 9 $\underline{> \text{ or } \geq}$ 3

Inequalities can be solved by finding values for the variable from a given set for which the inequality is true. Students determine which values make the inequality true by substituting each value in for the variable and evaluating the expression. If the inequality is true, then the value for the variable is a solution. Unlike equations, which have only one solution, inequalities can have many solutions.

Which of the numbers in the following set is a solution?

$f+2<9$; 6, 7, or 8

$6+2<9$
$8<9$ True

$7+2<9$
$9<9$ False

$8+2<9$
$10<9$ False

Out of the set of numbers {6, 7, 8}, 6 is the only solution to the inequality $f+2<9$.

$n-3>6$; 9, 10, or 11

$9-3>6$
$6>6$ False

$10-3>6$
$7>6$ True

$11-3>6$
$8>6$ True

Out of the set of numbers {9, 10, 11}, 10 and 11 are both solutions to the inequality $n-3>6$. In fact, any number greater than 9 will make this inequality true. We can graph this solution on a number line showing that n can be any number greater than 9.

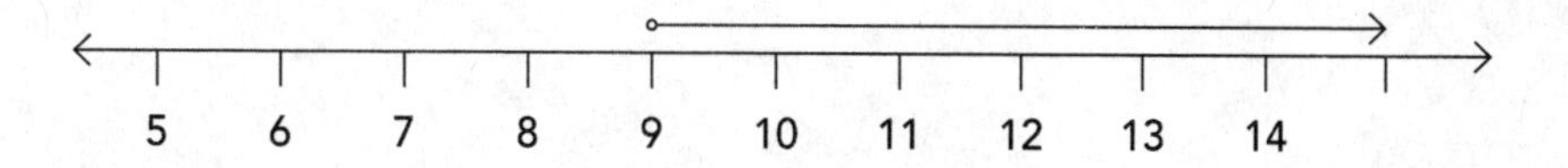

Inequalities are also used to represent scenarios when the answer can be a range of values, or infinitely many values. Instead of listing all possible solutions to the situation, an inequality is written and a number line diagram is created to show the solutions.

When drawing the number line diagrams, remember that a number line goes infinitely in both directions; however, it is impossible to draw such a line. Arrows on both ends of the number line are used to show that the number line extends from negative infinity to positive infinity.

There are different symbols that are used to model the inequalities. First and foremost, a ray is drawn on the number line to show which values are included as solutions. The endpoint of the ray has different meanings. If it is an open circle (not shaded in), this means that the value the ray is beginning on is not a solution to the set and is used with the inequality symbols less than

($<$) or greater than ($>$). If it is a closed circle (fully shaded in), this means that the value the ray is beginning on is a solution to the set and is used with the inequality symbols less than or equal to ($\leq$) or greater than or equal to ($\geq$).

Here are some examples:

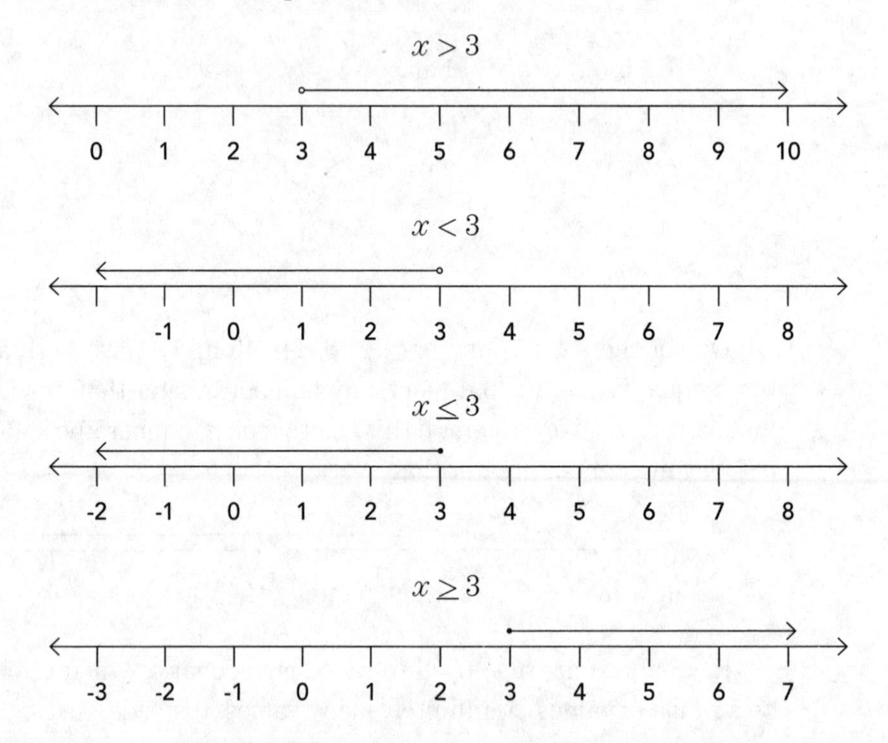

These are just the basics. Corestandards.org wants students to be able to write an inequality of the form $x > c$ or $x < c$ to represent a limitation or condition in a real-world mathematical problem, and to then be able to represent possible solutions on a number line. So, here we go.

How many of you have ever attended a carnival or amusement park and seen a sign that reads "You must be at least 12 years old to ride the go-karts." This is an inequality in the real-world setting. What do we do next?

The first step is to define the variable. What is the unknown in this statement? At this point, we do not know the age of the person wanting to ride the go-karts, so define a to be the age of the rider.

Next, we need to write an algebraic inequality using the appropriate symbol. How do we know what symbol to use? The following table is a list of commonly used words and the associated symbols.

Symbols	$<$	$>$	$\leq$	$\geq$
Words	Is less than, Is fewer than	Is greater than, Is more then	Is less than or equal to, Is at most, Is no more than, Maximum	Is greater than or equal to, Is at least, Is no less than, Minimum

Since this example says you must be at least 12 years old, you will need to use the greater than or equal symbol to write the inequality.

$$a \geq 12$$

Now you graph the possible solutions on a number line. How could you graph this? How much of the number line should you draw? It is recommended that you show at least a few numbers above and below the solution to the inequality, as it gives the reader a better picture. In this case I would show a number line beginning at 7 and extending until about 17, so that 12 is in the middle of the numbers. This is just one person's opinion; feel free to draw ten numbers above or below, or as many as you'd like.

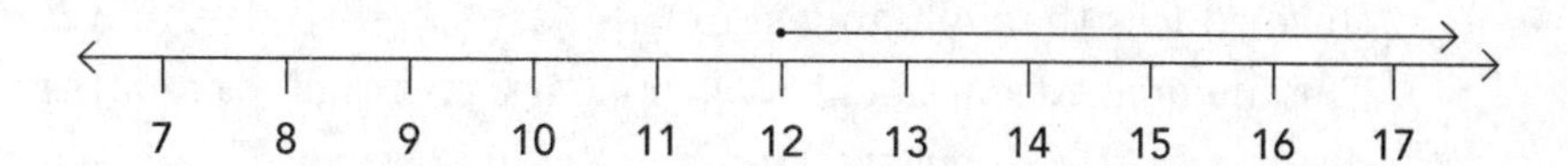

The last step is to check your hard work by testing a possible solution. Why is there a closed circle on 12? Is 12 greater than or equal to 12? The answer is yes, so 12 is a solution to the inequality. Is it true that a number to the right of 12, such as 14, will be a solution? Let's check. Is 14 greater than or equal to 12? Yes, so 14 is a solution. Let's check a number to the left of 12. Is 9 greater than or equal to 12? No, so 9 is not a solution and we have graphed the solution to the inequality correctly.

The maximum speed limit is 65 mph.

Let s = rate of speed. The inequality that represents this situation is $s <$ 65 mph. Now you try to make the graph on your own.

Your child may also be given a graph of an inequality and be asked to generate a real-world situation about what is represented. This is where he will take ownership for connecting the math he is learning in the classroom to the math that surrounds him every day.

For example, if given the following graph, a variety of real-world scenarios could be developed to match this inequality.

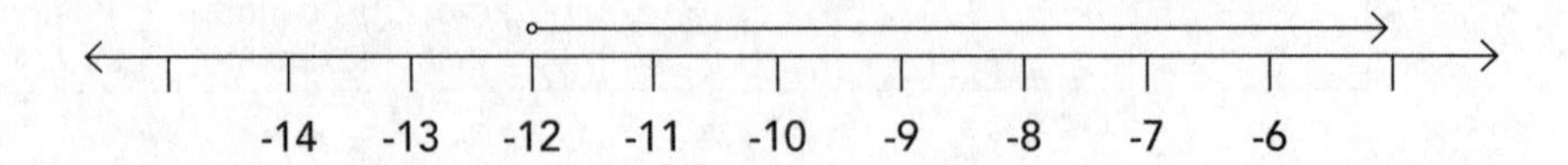

Here are some examples of real-world scenarios:

- The temperature last night was greater than 12 degrees below 0.
- The diver cannot go under 12 feet below sea level.
- The fewest points I had in *Jeopardy!* last night was −12 points.

Can you come up with others?

Sometimes you are faced with two limiting factors. For example, you cannot travel a negative distance. So if you were writing an inequality to represent the amount of distance traveled in a day, you might say that you traveled less than 400 miles. In this case, d = distance, and $d < 400$ miles. This inequality statement as it appears says that you can travel negative distances as well, but we know that is impossible. This is where you use a compound inequality. Since you know that the distance traveled for the day is between 0 and 400 miles, you use the following inequality. $0 \leq d < 400$. This can be drawn on a number line, and looks like a line segment rather than a ray.

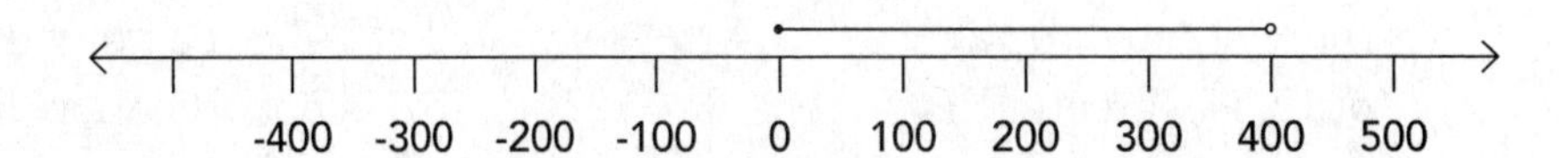

The most difficult aspect of inequalities and real-world situations is deciding which inequality symbol is associated with the words in the real-world situation. If your child is struggling with which symbol to pick, encourage him to put a number in for the variable that makes the scenario true and decide which symbol needs to be there to make the inequality true. Over time, the process will get easier.

Dependent and Independent Variables

Variables are used to represent unknown quantities in the real world. Oftentimes these variables are directly related to another quantity. This sets up a situation in which you have dependent and independent variables.

An *independent variable* is pretty true to its name: This is a variable that is not impacted or changed by another variable. It stands alone and isn't changed by anything else you are trying to measure. In fact, an independent variable is the variable that has an impact on others, more specifically the dependent variable. The independent variable is the input.

A *dependent variable* is true to its name as well: This variable depends on other factors and is changed due to these outside factors. Due to these reasons, the dependent variable is the output.

Let's use the amount of study time your child puts in preparing for a test. There are two variables in this situation. The first variable is the amount of time spent studying. The second variable is the score on the test. Does the score on the test impact the amount of time spent studying, or does the amount of time spent studying affect the score on the test? It is impossible for the test score to cause a change in the amount of time studying, since that happens before a test. So in this case, the amount of time spent studying is the independent variable, and the score on the test is the dependent variable.

Which is the independent and dependent variable in each of the following scenarios?

- Distance (d) and time (t)
  Since distance is calculated by rate • time, time is the independent variable and distance is the dependent variable. The amount of distance traveled depends on the amount of time spent traveling at a certain rate of speed.

- Amount of sunlight (s) and amount of plant growth (g)
  Since the amount a plant grows is dependent on the amount of sunlight it gets, the amount of sunlight is the independent variable and the amount of plant growth is the dependent variable.

- Heart rate (h) and amount of caffeine consumed (c)

The amount of caffeine consumed impacts your heart rate, so the amount of caffeine consumed is the independent variable and your heart rate is the dependent variable.

Once again, you can come up with others; just remember that your input is your independent variable, and your output is your dependent variable.

Function Tables

A function is a relationship that assigns exactly one output value to each input value. Function tables are a way to organize input and output values. A function rule describes the relationship between each input and output.

Let's look at a situation where building a function table helps you to better understand the problem.

> A local band charges $70 for each hour it performs. Define a variable. Then write a function rule that relates the total charge to the number of hours it performs. What is the cost of having a band perform for 5 hours?

To answer this question, first you must decide which is the independent and dependent variable. The two variables you have to consider are: number of hours the band performs, and total cost. Since the total cost is based on the number of hours the band performs, c (the total cost) is the dependent variable, and h (the number of hours the band performs) is the independent variable. The function rule is $70h = c$, because the rate per hour is $70.00.

▼ SAMPLE FUNCTION TABLE

Input	Output
Hours	**Total Cost**
h	$70h$
0	0
1	70
2	140
3	210
4	280
5	350
8	560

Based on the function table, it will cost $350 for the band to perform for 5 hours.

Another layer to this standard is that students can take the table of input and output values and turn them into ordered pairs that can be graphed on a coordinate plane. If a student wanted to analyze the relationship between hours performed and total cost of two different bands, a graph of these two relationships would help with this process. The steeper the line, the more expensive the band is per hour. The more gradual line is in fact the better deal. Although total cost is not the only contributing factor in the decision-making process, if you are working within a budget it can make a world of difference.

Practice Problems

Evaluate each of the following.

1. 5^2
2. $2^5 + 3^2 \bullet 5^0$
3. $6^1 \bullet 2^2$

Write each expression using exponents.

4. $2 \bullet 2 \bullet 3 \bullet 3 \bullet 3 + 4 \bullet 4 - 5$
5. $3 \bullet 3 \bullet 3 \bullet 1$

Write each verbal phrase as an algebraic expression.

6. Five more than six times a number.
7. A number less two.
8. Four times the difference of a number and three.
9. Nine less than the quotient of a number and seven.

Evaluate each of the following expressions.

10. If $n = 6$ and $m = 2$, what is the value of $nm - 4m + 16$?
11. If $k = 10$ and $j = 4$, what is the value of $2j + 8k \div 5$?
12. If $r = 6$, $s = 7$, and $t = 3$, what is the value of $rs \div (ts)$?
13. If $r = 6$, $s = 7$, and $t = 3$, what is the value of $rs \div ts$?

Identify the property being used.

14. $2+3+5=2+5+3$
15. $(2 \bullet 7) \bullet 4=2 \bullet (7 \bullet 4)$
16. $(6+3) \times 2=2 \times (6+3)$
17. $5 \times 1=5$
18. $6(3x+2y)=18x+12y$
19. $5(4b+3c)=5(3c+4b)$

Which value from the set is a solution to the given equation?

20. $3x+2=14$ {2, 3, 4, 5}
21. $7y-11=59$ {9, 10, 11}
22. $24=4p$ {6, 7, 8}

Solve for the variable in each equation.

23. $g+12=28$
24. $36=h-4$
25. $8k=72$
26. $d/6=12$

Write an inequality to represent each situation and then graph this on a number line.

27. The ice on the lake must be at least 4 inches thick to be safe to skate on.
28. The minimum score needed to pass the test is a 60%.
29. No more than 15 people can go on the roller coaster at a time.
30. The speed limit on the highway is 65 mph.

Create a function table for each of the following situations. Identify the independent and dependent variable.

31. An Internet company charges a one-time fee of $10 a year to be a member of its music program. It also charges $2 for each song you download. How much will it cost if you download 30 songs in a year?
32. You are traveling down a road at 35 mph. How far will you have traveled in 5 hours' time?

CHAPTER 8

Geometry

Geometry is the branch of mathematics that studies shapes, size, relative position of figures, and space. It focuses on all dimensions of mathematics, from the zero dimension (which is a point on a plane), the first dimension measuring length, the second dimension measuring area, to the third dimension measuring volume . . . and yes, there are more dimensions in mathematics, but that is for a whole different book. The focus of this chapter will be on area, surface area, and volume.

What Should My Child Already Be Able to Do?

The study of geometry begins as soon as your child is in preschool and learning to identify the most basic of shapes: square, triangle, circle, oval, rectangle, and star. They learn the early stages of classification; a shape with three sides is a triangle, a shape without sides and a round edge is a circle. Further development of the skills associated with identifying and describing shapes happens in kindergarten, where students learn to distinguish between two-dimensional "flat" shapes and three-dimensional "solid" shapes. They even begin to use such complex words as how many vertices (corners) a shape has. In first grade, students are using multiple two-dimensional shapes to create a composite shape, and then make new shapes from that. The study of area begins informally in second grade, where students break a rectangle into rows and columns of same-sized squares and count to find the total number of them.

In third grade, students learn about area and perimeter. They continue to learn about area by counting the number of square units inside the plane figure (two-dimensional figure) and expand upon this understanding to show that the area of a rectangle can be found by tiling it, or by multiplying the side lengths with the formula $Area = length \times width$. Next they move on to perimeter, learning that perimeter is the sum of all of the side lengths and that it measures the distance around the shape, while area measures the space inside the shape.

Let's fast-forward to fifth grade, where students learn that volume is an attribute of a three-dimensional figure. They also learn that volume is the idea of packing a solid figure (cube, rectangular prism, etc.) with unit cubes, 1 unit by 1 unit by 1 unit, without gaps or overlaps. The number of unit cubes used to completely fill the solid is called the *volume*. They progress from counting the number of unit cubes used to applying the formula of $V = l \times w \times h$. They continue their learning to see that volume is additive, meaning if you have a figure that is composed of two right rectangular prisms (prisms with faces that are perpendicular to the base), you can find the volume of each individually, and add the volume of the non-overlapping parts together to find the total volume of the figure itself.

So by the time they reach sixth grade they should have learned how to determine the area and perimeter of rectangles using whole number side lengths, calculate the volume of rectangular prisms, and should know the corresponding formulas. Your job is to take it to the next step.

Area

Area is the study of the size of a surface. Many shapes can have the same area, but have very different configurations. Each of the figures drawn here has an area of 10 square units, but they all look very different.

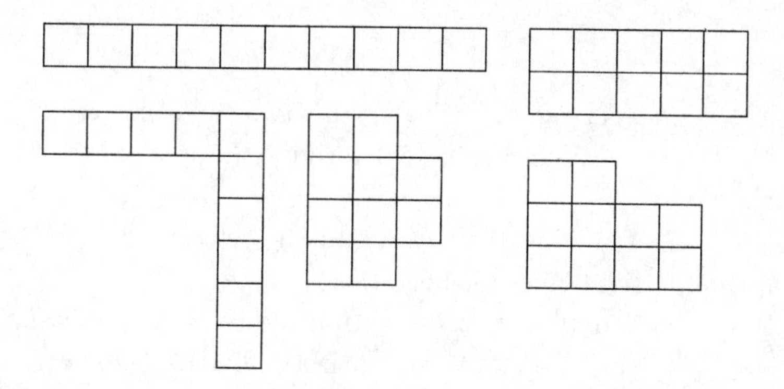

Learning how to calculate area is an important skill, and one that is used frequently. Area can be used to figure out how many gallons of paint you need to paint the walls of your home, how much grass seed is needed to plant a lawn, how many tiles you need to tile your bathroom floor, or carpet for your bedrooms. Area is everywhere. Carpenters use calculations of area every day in their work. Architects use area to determine the square footage of a home. Ideally, all of the shapes you would calculate area for would be rectangles, but we know this is not true.

In sixth grade, students use their knowledge of area of rectangles to find the area of right triangles, other triangles, special quadrilaterals, and polygons by learning to compose (put things together) and decompose (take them apart) them into shapes that are familiar. From there, students work to develop area formulas of more specific shapes.

From Rectangles to Triangles

To continue to support the eight Standards for Mathematical Practice as set out in the Common Core Standards, teachers create activities that allow students to develop the formula for area of a triangle through guided exploration. A teacher may engage students in a similar process being shared with you here.

What happens when you take a rectangle and cut it in half along its diagonal?

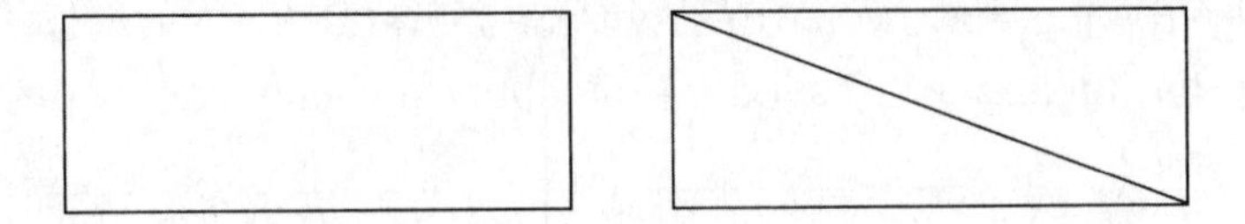

The rectangle is broken up into two equal-sized right triangles. How can you use this information to determine the area of one of the right triangles?

1. You know that the area of a rectangle is found by multiplying the base times the height, or $A = b \bullet h$.
2. You know that by cutting the rectangle in half along its diagonal, you created two equal-sized right triangles.

By pulling these two pieces of information together, you learn that the area of a right triangle is equal to $\frac{1}{2}$ of the area of the rectangle, or by using the following formula:

$$\textit{Area of a Right Triangle} = \frac{\textit{base} \bullet \textit{height}}{2}$$

But this is only for a very special type of triangle, the right triangle. What would happen if you had an isosceles triangle; what would the area be then? A drawing of an isosceles triangle is pictured on the left. That same triangle is then placed inside a rectangle with the same base and height as the triangle. Do you notice anything?

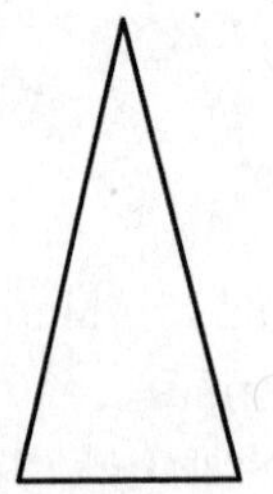 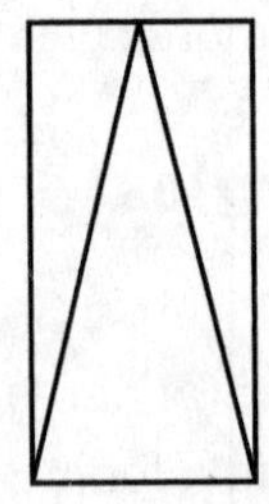 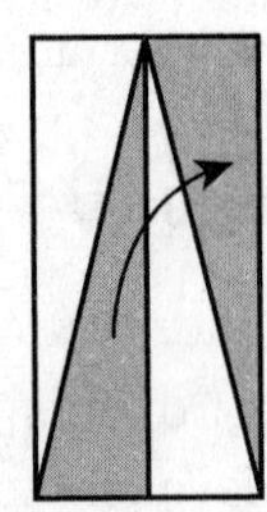 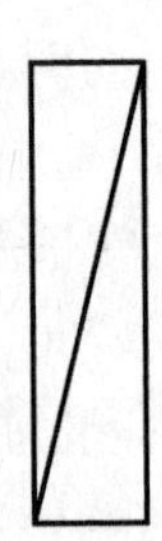

Look at the third image. Notice how by shading one half of the triangle, and flipping that shaded piece, we are able to position it inside the rectangle in such a way that the two pieces together form a rectangle one-half the size of the original rectangle. (Hmm, that is interesting.) So does the formula for the area of a right triangle work for all triangles? Yes, it does!

The height of a triangle, also known as the *altitude*, is found by drawing a perpendicular line from the base of the triangle to the opposite vertex. Depending on the configuration of the triangle, it is sometimes necessary to extend the base to draw the altitude. Given there are three possible bases, there are also three different altitudes. For our purposes, the base can be any of the three sides, but is usually the one drawn at the bottom.

Students can be given a variety of styles of problems to calculate area of a triangle. Some problems may have a drawing and students need to pull key information from the drawing; others may be given the base and height of the triangle and asked to find the area that way. Triangles may even be part of a more complex figure composed of several shapes. Regardless of how the problem is presented, the formula for area of a triangle holds true. $Area\ of\ a\ triangle = base \times height \div 2$.

A triangle has a base of 8 cm and a height of 5 cm. What is the area?

When solving a problem like this, you should be sure to show all of your work, beginning with the formula first.

$Area = \frac{b \bullet h}{2}$ Begin with formula first.

$Area = \frac{8\text{ cm} \bullet 5\text{ cm}}{2}$ Substitute the values for the base and height.

$Area = \frac{40\text{ cm}^2}{2}$ Simplify the numerator.

$Area = 20\text{ cm}^2$ Simplify the fraction.

Notice that each step of the work was shown and the final answer had units of measurement squared. Since area is a two-dimensional measurement, the label must include square units as well.

Rectangles and Parallelograms

If students remember anything from classifying quadrilaterals, they should know that every rectangle is a parallelogram (but not every parallelogram is a rectangle). With that being said, a parallelogram can be turned into a rectangle with some very simple cutting and pasting. In the following image of the parallelogram, a perpendicular line has been drawn showing the height. The area to the left of that line has been shaded, forming a triangle. Imagine you could cut this parallelogram along the line representing its height, and slide it to the right edge. What would be formed?

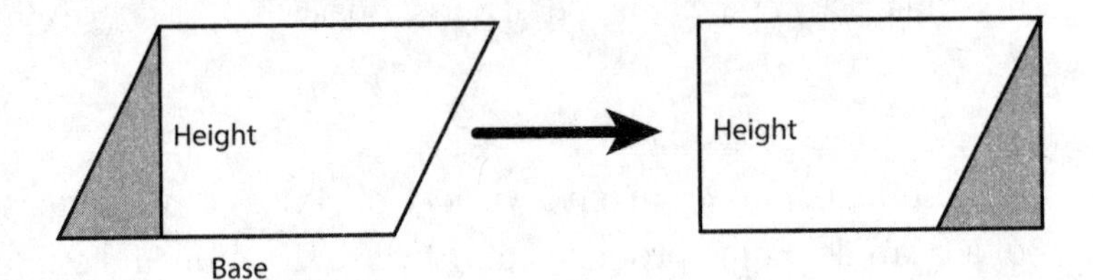

Hopefully you said a rectangle would be formed. If you did, you are right! So what is the formula for determining the area of a parallelogram? It is the exact formula as that of a rectangle.

$$Area\ of\ a\ parallelogram = base \times height$$

Remember, no matter what shape you are working with, the height always has to be perpendicular to the base. If it is at an angle (meaning not perpendicular), that measurement cannot be used as the height in the formula.

What is the area of a parallelogram with a length of 2.5 cm and a height of 1.25 cm?
$Area = b \bullet h$
$Area = 2.5\text{ cm} \bullet 1.25\text{ cm}$
$Area = 3.125\text{ cm}^2$

Notice that when showing the work, each step is a new line. Requiring your child to show his work in this fashion will make every teacher's dream come true. A teacher loves it when students take pride in their work and show organized, well-thought-out answers!

Area of a Trapezoid

A trapezoid comes with a lot of controversy surrounding its definition. Some say a trapezoid is a quadrilateral with at least one pair of parallel sides. By this definition, all parallelograms are trapezoids. Others define a trapezoid to be a quadrilateral with exactly one pair of parallel sides, meaning that no parallelogram is a trapezoid. For our purposes, we are going to define a trapezoid to be a quadrilateral with exactly one pair of parallel sides.

Based on what you have learned so far, you know the area formula of a rectangle and parallelogram as base times height. You were also able to show that the area of a triangle is half that of a rectangle with the same base and height. Knowing that these shapes can be manipulated to form another more familiar shape helps to derive the area formulas. Let's try it with the trapezoid.

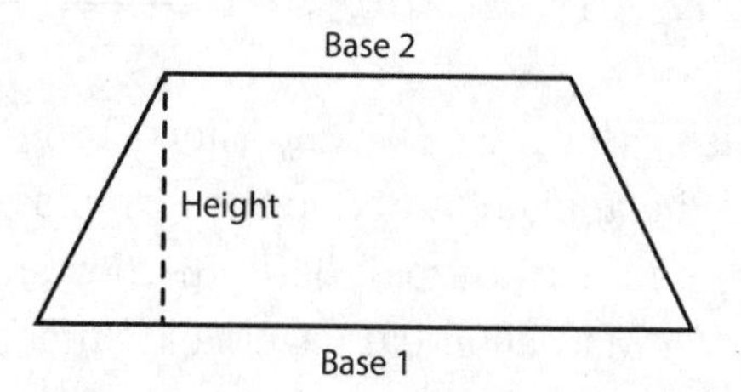

A trapezoid has exactly one pair of parallel lines that act as the bases; however, since they are different lengths, you name them base 1 and base 2. The trapezoid's height is a perpendicular line that extends from one base to the other. Is there a way that you can turn this into a more familiar shape? Can it become a rectangle, a triangle, a parallelogram with an easy manipulation of its existing shape? The answer is yes. If you duplicate the existing trapezoid and flip the second trapezoid over a horizontal line, you can turn it into a parallelogram. Look and see.

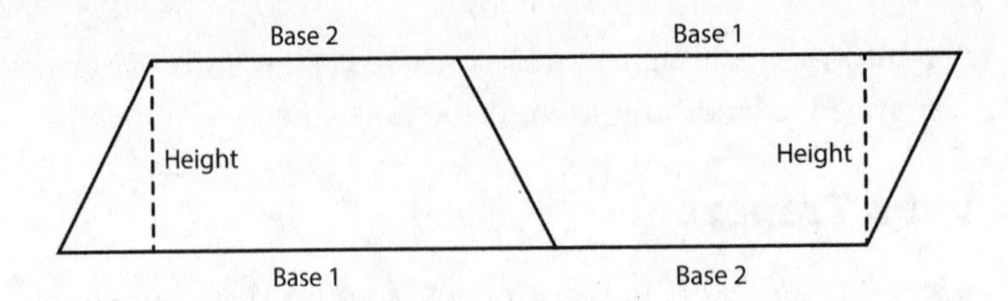

The parallelogram has been created using two identical trapezoids. What is the length of the base of the newly formed parallelogram? The length of the base is the sum of base 1 and base 2. Has the height changed? No. So let's pull this all together:

The area of a traditional parallelogram is $base \times height$.

The area of this parallelogram is $(base\ 1 + base\ 2) \times height$.

What is the area of the initial trapezoid? Since the parallelogram is made up of two trapezoids, then the area of one trapezoid must be $\frac{1}{2}$ of the area of the parallelogram.

$$Area\ of\ a\ trapezoid = \frac{(base\ 1 + base\ 2) \bullet height}{2}$$

Wow, isn't that neat? Using information you already knew about rectangles can be applied to derive the formula for the area of a parallelogram, a triangle, and a trapezoid. This further supports the goals of the eight Standards for Mathematical Practices. Students are extending their understanding and making sense of new mathematical concepts on their own. Even if they forget the formula, by having the conceptual understanding they may be successful in finding the area of any trapezoid.

Area of Polygons and Other Irregular Figures

Now that you know the area formula for triangles, rectangles, and trapezoids you can pretty much figure out any other figure by decomposing it into these familiar shapes. For example, a hexagon can be broken into two congruent trapezoids, or six congruent triangles. Once you find the area of one of the decomposed figures, you can multiply to find the total area. The key thing to remember is identifying which length is the height; it must always be perpendicular to the line segment you are using as your base.

Composite figures are exactly what the name says, figures that are made of up two or more simpler figures. Here are a few examples:

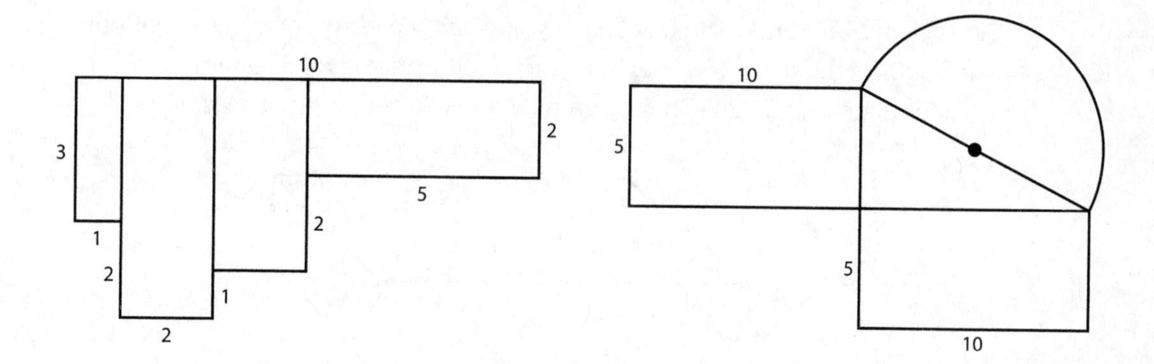

Notice how in both of these drawings, these irregular shapes were broken down into much more familiar-looking objects such as rectangles, triangles, half-circles, etc. It is now feasible to calculate the area.

Surface Area

Surface area is the total area of the surface of a three-dimensional object. Imagine you are wrapping a gift box for a friend's birthday. The surface area of the box is the amount of wrapping paper that is needed to fully cover the box. Think about a cereal box. Its surface area is found by measuring the area of all six of its faces. Now, surface area is not just found on rectangular prisms. You can find the surface area of a cube, a prism, a sphere, or a cylinder, simply by knowing the formula. But what if you don't have the formulas memorized? How else could you measure surface area?

Learning math is often like building a house: You need your foundation before you can build up. The same is true here. Being able to calculate surface area is dependent on being able to find the area of the most common two-dimensional shapes. This knowledge serves as your foundation. If students have not learned the formulas for area of rectangles, squares, triangles, circles, trapezoids, or other polygons, then taking the next step to learn about surface area is nearly impossible. Why, you ask? The answer is simple: Every three-dimensional solid is composed of two-dimensional figures.

A rectangular prism is a solid object, composed of six rectangular faces. It is named a rectangular prism because the two bases (parallel faces) are rectangular in shape. Take the cereal box, for instance; if you cut along every edge and broke the box apart, you would have six rectangles, but only three different sizes. You see, in a rectangular prism, parallel faces are congruent (same size). Let's look at an example of a rectangular prism.

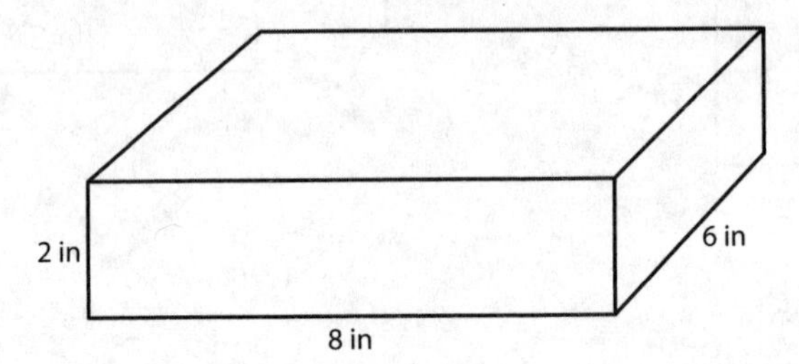

A net can be created from this rectangular prism. A *net* is an unfolded geometric solid. It is a two-dimensional sketch of all of the faces that form the geometric solid. Here a net has been made of the rectangular prism.

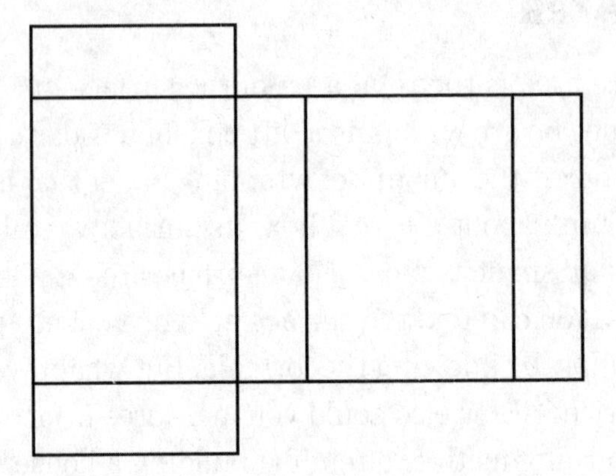

And now, when adding the square units you can determine the surface area one of two ways: by counting the squares, or by using the area formula for a rectangle to determine the area of each of the three different rectangles. Let's try it both ways.

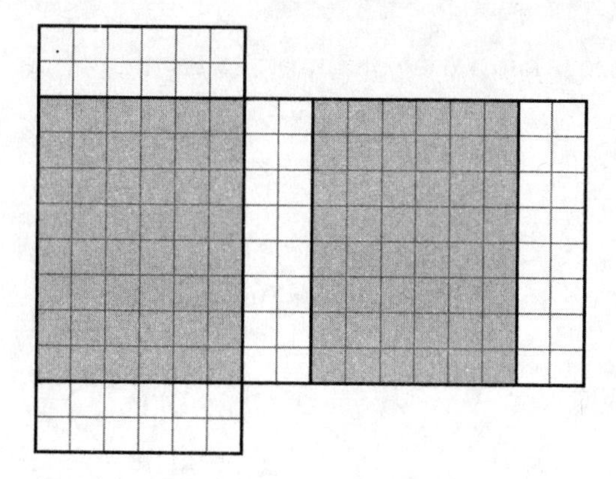

If you count the squares, the surface area of the rectangular prism is 152 square inches. If you compute the areas of the rectangles, you only need to do three calculations because each face has one congruent face, or a face with the exact dimensions.

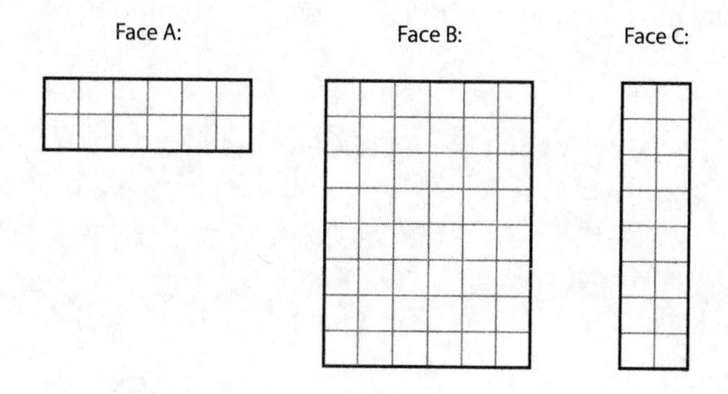

Face A has a base of 6 in and a height of 2 in.

$$\begin{aligned} Area &= b \times h \\ &= 6 \text{ in} \times 2 \text{ in} \\ &= 12 \text{ in}^2 \end{aligned}$$

Since there are two faces with these dimensions, the total area is 2×12 $\text{in}^2 = 24 \text{ in}^2$.

Face B has a base of 6 in and a height of 8 in.

$$\begin{aligned} Area &= b \times h \\ &= 6 \text{ in} \times 8 \text{ in} \\ &= 48 \text{ in}^2 \end{aligned}$$

Since there are two faces with these dimensions, $2 \times 48 \text{ in}^2 = 96 \text{ in}^2$.

Face C has a base of 2 in and a height of 8 in.

$$\begin{aligned} Area &= b \times h \\ &= 2\text{ in} \times 8\text{ in} \\ &= 16\text{ in}^2 \end{aligned}$$

Again, there are two faces that share these dimensions. $2 \times 16\text{ in}^2 = 32\text{ in}^2$.

The total surface area is found by adding the three individual areas together.

$$\begin{gathered} 24\text{ in}^2 + 96\text{ in}^2 + 32\text{ in}^2 \\ 120\text{ in}^2 + 32\text{ in}^2 \\ 152\text{ in}^2 \end{gathered}$$

Either method achieved the same calculation for surface area.

Now, these calculations were done with whole numbers. Surface area can be determined using any positive real number.

Why can't you use negative numbers?

Remember that length is a measure of distance, and distance is always positive. Even if you traveled backward 5 meters, you still traveled a distance of 5 meters.

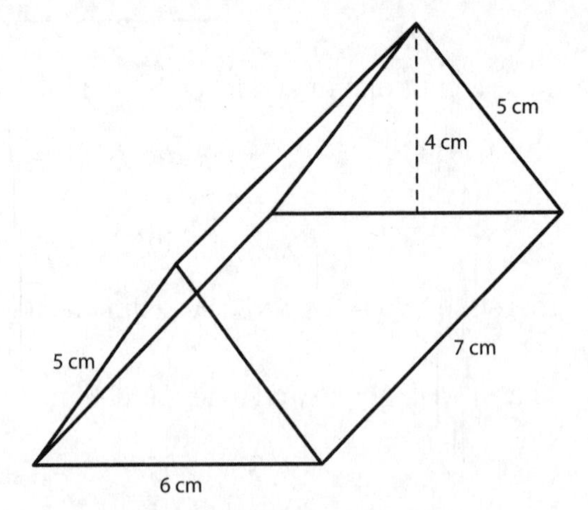

Look at the next figure; it is a triangular prism. It is named this because its two parallel faces are triangles and its sides are all parallelograms, or more specifically in this case, rectangles. If you look at the net, you will see each of these shapes.

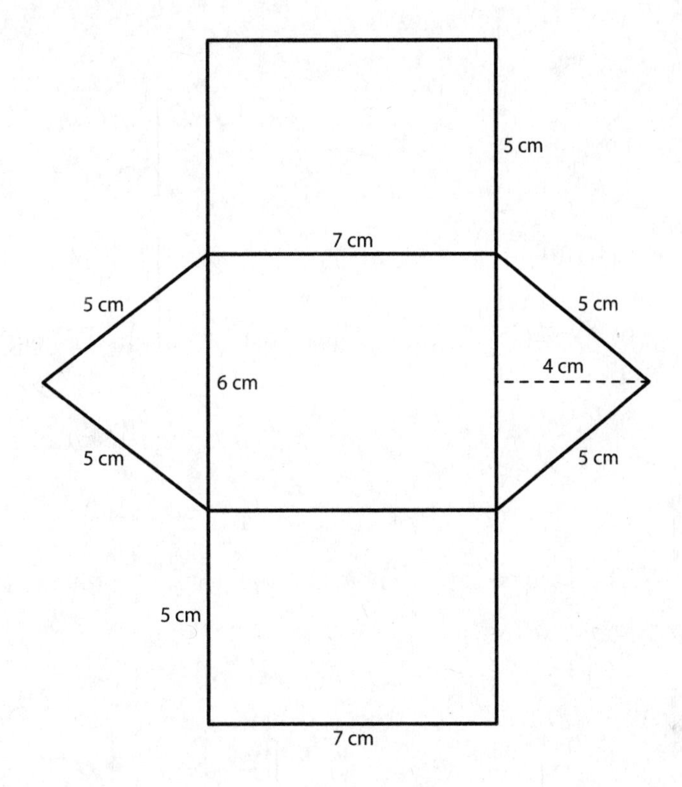

To figure out the surface area of this solid, you need to find the area of three different shapes:

1. The area of the triangles, which each have a base of 6 cm and a height of 4 cm.

$$Area\ of\ a\ triangle = \frac{b \bullet h}{2}$$

$$= \frac{6\text{ cm} \bullet 4\text{ cm}}{2}$$

$$= \frac{24\text{ cm}^2}{2}$$

$$= 12\text{ cm}^2$$

Since there are two triangles with this area, the total area of the triangles is 24 cm^2.

2. Two rectangles have the same dimensions, a base of 7 cm and a height of 5 cm.

$$\begin{aligned} \textit{Area of a rectangle} &= b \bullet h \\ &= 7\text{ cm} \bullet 5\text{ cm} \\ &= 35\text{ cm}^2 \end{aligned}$$

Since there are two of the rectangles with these dimensions, the total area is 70 cm^2.

3. The last rectangle has a base of 7 cm and a height of 6 cm.

$$\begin{aligned} \textit{Area of a rectangle} &= b \bullet h \\ &= 7\text{ cm} \bullet 6\text{ cm} \\ &= 42\text{ cm}^2 \end{aligned}$$

The surface area of the triangular prism is equal to the sum of the three areas.

$$12\text{ cm}^2 + 70\text{ cm}^2 + 42\text{ cm}^2$$
$$82\text{ cm}^2 + 42\text{ cm}^2$$
$$124\text{ cm}^2$$

These are just two examples of how to determine surface area. The value of knowing and understanding surface area plays a role in product development and design. Manufacturers want to build boxes to hold their goods with the least amount of production cost. Decreasing surface area requires the use of less materials and decreases the overall cost of production. This is a great investigation for students to complete while learning about surface area and volume.

Volume

Volume measures how much space a three-dimensional solid takes up. It also is a measure of a solid's capacity, how much it can hold. Let's use the previous rectangular prism. How much space does this prism take up?

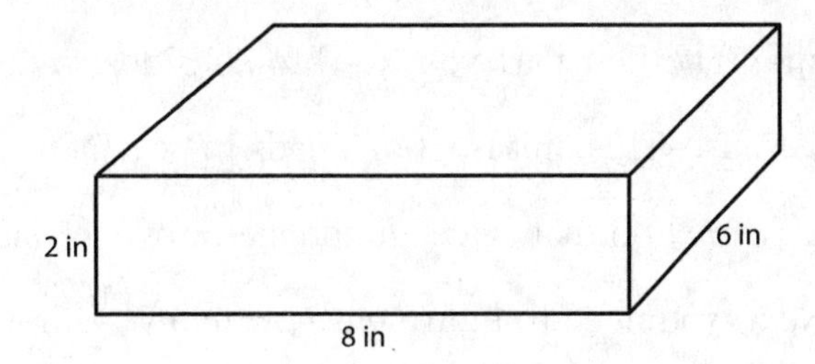

Volume is a three-dimensional measurement, found by multiplying the length times the width times the height.

$$Volume = l \times w \times h$$

Using the measurements of length, width, and height, the volume of the rectangular prism is:

$$\begin{aligned} V &= l \times w \times h \\ &= 8 \text{ in} \times 6 \text{ in} \times 2 \text{ in} \\ &= 48 \text{ in}^2 \times 2 \text{ in} \\ &= 96 \text{ in}^3 \end{aligned}$$

Not so bad, right? Although this is how to calculate volume, this type of problem would be mastered in the fifth grade. Corestandards.org expects sixth-graders to "find the volume of a right rectangular prism with fractional edge lengths by packing it with unit cubes of the appropriate unit fraction edge lengths, and show that the volume is the same as would be found by multiplying the edge lengths of the prism." This is conceptually more difficult because students don't think of things in terms of fractional parts of a number.

Let's break the standard down: *Fractional edge lengths* means that a side of the prism measures $3\frac{4}{5}$ feet, for example. You can handle that. It is the idea of packing it with unit cubes of the appropriate unit fraction that can be confusing. A *unit fraction* is a fraction where the numerator is always 1, and the denominator is any positive integer, $1/n$, where $n > 0$. Several examples of unit fractions would be $\frac{1}{2}, \frac{1}{3}, \frac{1}{5}, \frac{1}{10}$, etc.

Now that you have the vocabulary sorted out, let's think about the packing of the rectangular prism with fractional side lengths. If a rectangular prism is $2\frac{1}{2}$ units long, by 3 units wide, by $4\frac{1}{2}$ units tall, what is its volume? The first part you need to consider is what unit fraction are you going to use. This unit fraction needs to be a factor of each side length. $\frac{1}{2}$ is the unit fraction that is a common factor of each side length.

Next, you need to figure out how many $\frac{1}{2}$-unit cubes you will need for each side.

$2\frac{1}{2}$ units long $= \frac{1}{2} + \frac{1}{2} + \frac{1}{2} + \frac{1}{2} + \frac{1}{2}$, or 5 cubes long

3 units wide $= \frac{1}{2} + \frac{1}{2} + \frac{1}{2} + \frac{1}{2} + \frac{1}{2} + \frac{1}{2}$, or 6 cubes wide

$4\frac{1}{2}$ units tall $= \frac{1}{2} + \frac{1}{2} + \frac{1}{2} + \frac{1}{2} + \frac{1}{2} + \frac{1}{2} + \frac{1}{2} + \frac{1}{2} + \frac{1}{2}$, or 9 cubes tall

If you were to draw this using the $\frac{1}{2}$-unit cubes, you would need it to be 5 cubes long, 6 cubes wide, and 9 cubes tall. How many $\frac{1}{2}$-unit cubes are needed to build this rectangular prism?

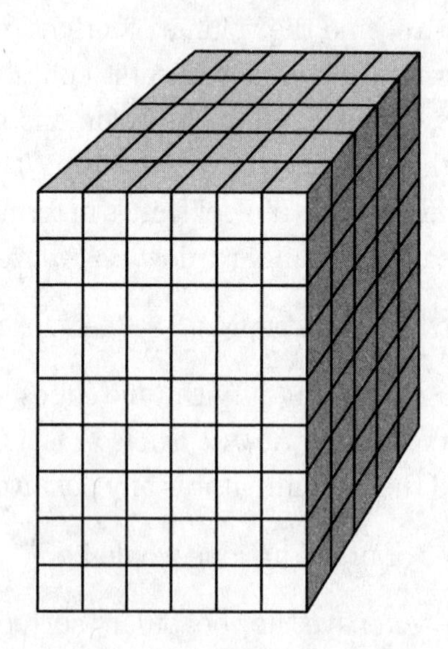

Again, the number of cubes needed could be found by counting them each individually, or you could find the number of cubes needed for the bottom row and use repeated addition for the number of rows needed to create this prism.

The bottom row of the prism requires 5 cubes by 6 cubes, or a total of 30 cubes, to make the first row. There are nine total rows. Using repeated addition we can determine the total number of cubes in the figure.

$$30+30+30+30+30+30+30+30+30=270 \text{ cubes}$$

270 represents the number of $\frac{1}{2}$-unit cubes needed to create this shape, but it is not the volume in whole units. How many $\frac{1}{2}$-unit cubes does it take to create 1 cubic unit (1 unit by 1 unit by 1 unit)?

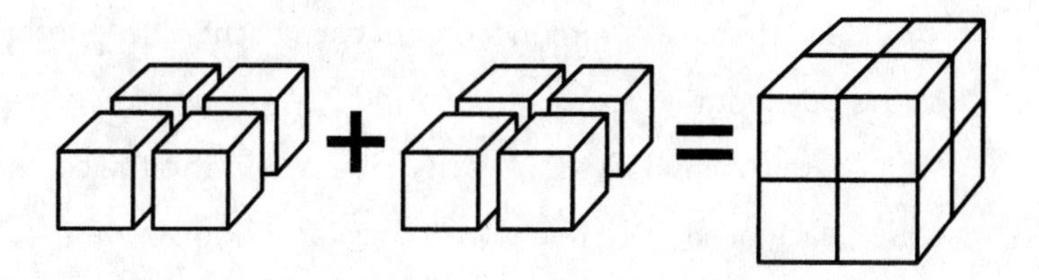

Two $\frac{1}{2}$-unit cubes are needed to make 1 whole unit, and a cube is composed of three dimensions, so 2 cubes by 2 cubes by 2 cubes is 8 total cubes needed for 1 cubic unit. You know that there are 270 $\frac{1}{2}$-unit cubes in the drawing, so if you take 270 and divide by 8 (the number of $\frac{1}{2}$-unit cubes in a cubic unit) you will be able to determine the volume. Using traditional division we can determine the quotient of 270 and 8.

$$\begin{array}{r} 33.75 \\ 8\overline{)270.00} \\ \underline{-24} \\ 30 \\ \underline{-24} \\ 60 \\ \underline{-56} \\ 40 \\ \underline{-40} \\ 0 \end{array}$$

The quotient of 270 divided by 8 is 33.75, or $33\ \frac{3}{4}$ cubic units.

How do you know what unit fraction to use to pack a rectangular prism of fractional side lengths?

You need to find a common factor of each of the side lengths. For instance, if the side lengths were $2\ \frac{1}{4}$, $5\ \frac{1}{2}$, and $2\frac{3}{4}$, the unit fraction you should use is $\frac{1}{4}$ because it is a factor of each number. It is also the common denominator of each of all three fractions. If the side lengths were $7\frac{2}{3}$, $2\frac{5}{6}$, and 4, the unit fraction you should use is $\frac{1}{6}$. Again, 6 is the common denominator of all three of the fractions.

You might be thinking that there has to be an easier way to figure out the volume of fractional unit lengths. And there is. You know that volume can be calculated using $l \times w \times h$. You also know how to multiply fractions and mixed numbers. To calculate the volume of rectangular prisms with fractional length sides, you put these two ideas together.

Using the previous example, determine the volume. If a rectangular prism is $2\frac{1}{2}$ units long, by 3 units wide, by $4\frac{1}{2}$ units tall, what is its volume?

$V = l \times w \times h$

$= 2\frac{1}{2}$ units $\times$ 3 units $\times$ $4\frac{1}{2}$ units — Change mixed numbers into improper fractions.

$= \frac{5}{2}$ units $\times$ $\frac{3}{1}$ units $\times$ $\frac{9}{2}$ units — Multiply numerators and denominators.

$= \frac{15}{2}$ units^2 $\times$ $\frac{9}{2}$ units

$= \frac{135}{4}$ units^3 Simplify the fraction.

$= 33\ \frac{3}{4}$ units^3

Although this method seems a lot easier, it doesn't focus on the conceptual understanding of what it means to have a fractional length.

Combining the conceptual understanding with the procedural understanding provides students with two avenues with which to approach the problem. Some students are better as visual learners, others are better with calculations. Let your child figure out which way makes most sense to him.

Practice Problems

Calculate the area.

1. What is the area of a parallelogram with a base of 4.2 in and a height of 3.5 in?
2. What is the area of a triangle with a base of 9.5 m and a height of 6.5 m?
3. What is the area of a trapezoid with the following dimensions: base 1 = 12 ft, base 2 = 7 feet, height = 5 ft?
4. Determine the area of the composite figure pictured here.

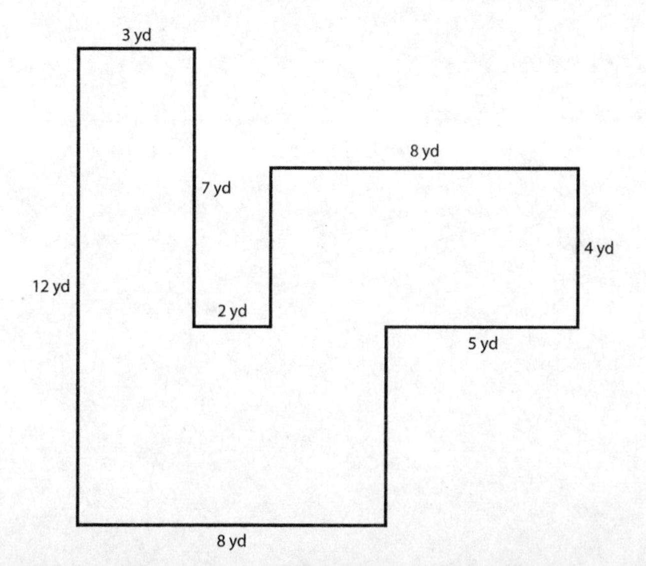

Answer the following questions about surface area and nets.

5. A rectangular prism has a length of 2 ft, a width of 3 ft, and a height of 6 ft. What is the surface area?
6. Draw a net for a rectangular prism that has dimensions 6 cm by 8 cm by 13 cm, and then determine the surface area.
7. A triangular prism has a base of 5 m, an altitude of 12 m, and an overall height of 15 m. What is the surface area?

Calculate the volume.

8. How many $\frac{1}{4}$-unit cubes does it take to make a rectangular prism that is $3\frac{1}{2}$ units by $5\frac{3}{4}$ units by $2\frac{1}{4}$ units in size?
9. What is the volume of the rectangular prism described in previous question?
10. What is the volume of a cube that is $1\frac{1}{2}$ inches long on each side?
11. Tina rented a storage unit in the shape of a rectangular prism. The storage unit has a width of $7\frac{3}{4}$ ft, a length of $12\frac{1}{2}$ ft, and an overall volume of $968\frac{3}{4}$ ft^3. What is the height of the storage unit?

CHAPTER 9

Statistics and Probability

In sixth grade, students learn about what types of questions make statistical questions and why. They learn to collect data, organize it into displays, and use measures of center to summarize data. They also learn that measures of variation describe how the values in the data set vary. Students are often quite comfortable with calculating the mean, median, and mode, so they feel comfortable with the start of this unit.

What Should My Child Already Be Able to Do?

Students begin learning about data at a very early age. It begins with simple data collection and the answering of questions, such as what is the total number of data points, how many are in each category, and how many or less are in one category versus another. In grade 2, students move toward creating line plots, picture graphs, and bar graphs to represent a data set of no more than four categories. In fourth and fifth grade, students are able to create line plots displaying data reflecting measurements to the nearest $\frac{1}{8}$ of a unit. It is not until sixth grade that students learn about measures of center, including mean, median, and mode.

What Makes a Statistical Question?

There are two types of questions, statistical questions and non-statistical questions. A *statistical question* is answered by collecting a set of data. A statistical question will generate a variety of answers. To answer a statistical question, you need to be able to collect data and analyze it. The data collected may be either categorical or numerical, depending on the question being asked. An example of a categorical question is, "What is your favorite animal?" Answers to this question will be categories of animals, such as dog, cat, horse, monkey, or elephant. An example of a numerical question is, "How many siblings does each member of our class have?" Students would respond with numbers: I have 1 brother, I have 2 siblings, I have no siblings. A *non-statistical question* has a single answer; there is no variability in the response. An example of this would be, "How old are you?" A person can only be one age. Let's look at a few examples in the following table.

Statistical Questions	Non-Statistical Questions
How old are the students attending the middle school?	How old are you?
How long do students spend on math homework?	How long did you spend on homework today?
How many siblings do each of your classmates have?	How many siblings do you have?
Which day this month had the highest temperature?	What is the temperature today?
Do dogs run faster than cats?	Is my dog faster than your cat?
Do firefighters make more money than police officers?	Do I make more money than you?

In sixth grade, the primary focus should be on statistical questions that generate numerical data. Once the data is collected, students will learn how to organize it into a table or display it in a graph.

How Can Data Be Described?

Sixth-grade students will use two ways to summarize a data set. Data that answers a statistical question can be described using *measures of center*, also known as measures of central tendency. They also can be summarized by *measures of variation*, also called measures of spread or measures of dispersion. Regardless of which choice you make, it often helps to make a graph of the data first.

Measures of Center

Measures of center describe how data looks at the center. A single number is used to summarize all the values. These are probably more commonly known to you as the measures of central tendency, or mean, median, and mode. Students in sixth grade will spend the majority of their time utilizing mean and median to describe data; however, it doesn't hurt to know about the mode.

Mean

The *mean* is also known as the average. It is the sum of all of the values in the data set divided by the number of values in the set. It equally distributes the data over all the numbers in the set. There are two ways to determine the mean: the algorithm used to calculate mean, and by equally distributing picture objects.

Let's explore the idea of using pictures as a way to determine the mean.

Kayla, Maya, John, Paul, and Christine went trick-or-treating and agreed to share their collected candy evenly. How much did each person get?

The following table shows how much candy each person collected.

Name	Amount of Candy Collected
Kayla	7
Maya	4
John	9
Paul	8
Christine	12

What might this look like as a picture?

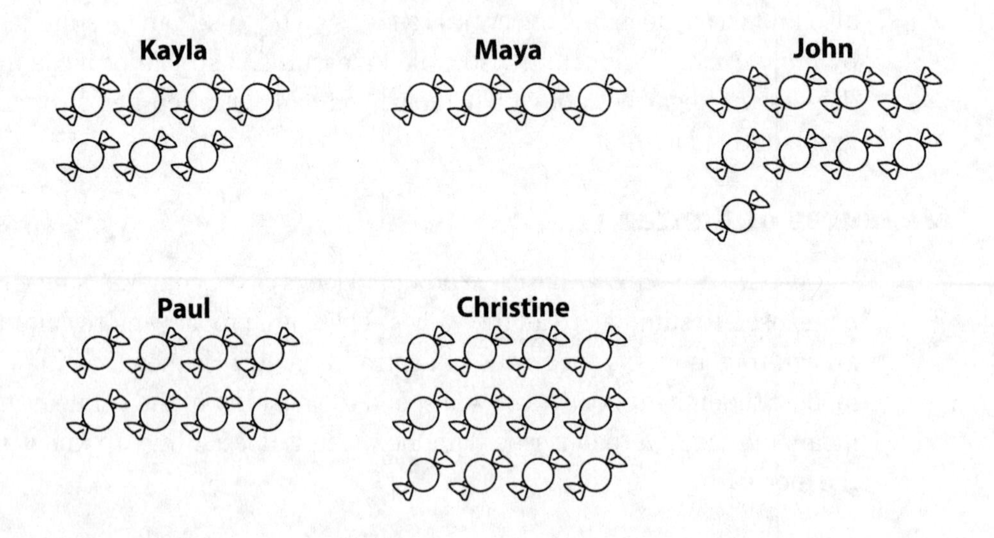

Now, let's put all the pieces of candy into one pile. This is known as combining the data.

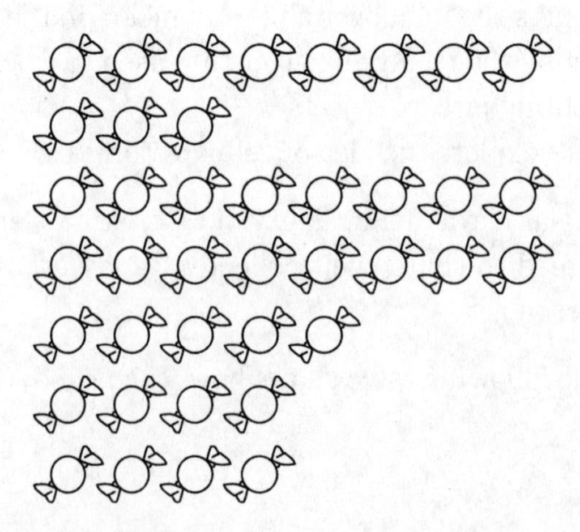

To find the sum, you can count all the pieces of candy in the circle, or you can organize it into an array and use multiplication to find the total number of pieces of candy.

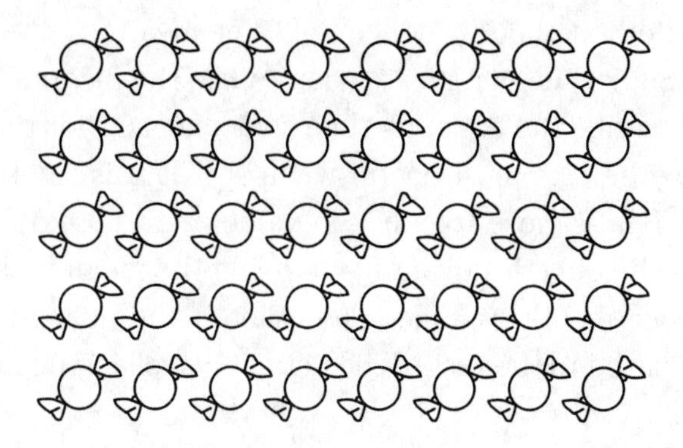

Each row in the array contains 8 candies, and each column contains 5. Using multiplication, $8 \times 5 = 40$, we can determine there are 40 candies. Now, this idea of equal distribution requires you to deal out the candy to each person, much like you are dealing cards for a card game, so that everyone has the same amount. Leveling out the data or making each set of data equal determines the mean.

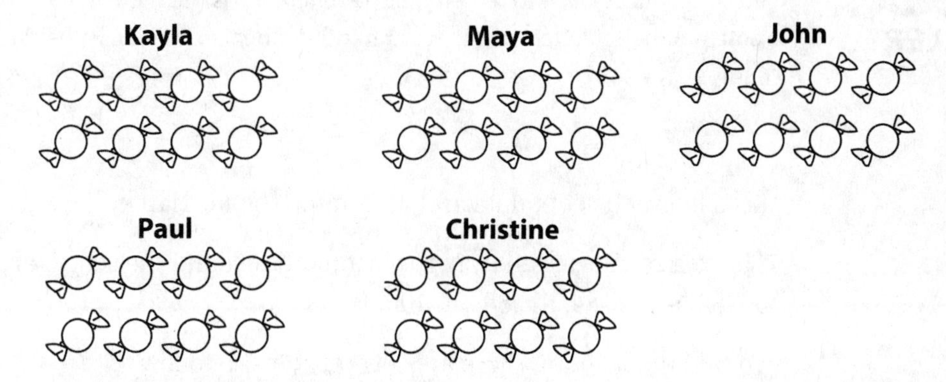

The mean for this set of data is 8 and was found by equally distributing the objects.

Let's see if we arrive at the same answer by using the algorithm. The numerical data is 7, 4, 9, 8, and 12. The algorithm says to find the sum of the data and then divide by the number of pieces of data in the set. There are 5 pieces of data in the set.

$$\begin{aligned} \text{Mean} &= (7+4+9+8+12) \div 5 \\ &= 40 \div 5 \\ &= 8 \end{aligned}$$

Both methods produce the same mean.

So how do you know when to use the mean? The mean should be used to describe the data any time someone asks for the average of the data set. However, it is most useful when the data is tightly clustered, meaning all the data is close together in value, with no extreme data points or outliers. The only disadvantage to using the mean is that it is vulnerable to the effects of outliers. Since an outlier is an unusually high or low value when compared to the rest of the data, it can skew the mean.

Median

The *median* is best described as the middle number of a set of values, when the numerical data is ordered from least to greatest. If there are two middle numbers, then the median is the mean of those two numbers.

Students often confuse mean and median because they both start with the letter m. An easy way to remember the difference is the median has the same number of letters as the word 'middle'—hence, median is the middle number in a set.

Let's look at a set of data and determine the median.

The test results are in from the last math test and the scores are: 95, 56, 82, 97, 95, 84, 92, 88, 91, 100.

To find the median, there are several steps to follow.

1. First, you need to arrange the data from least to greatest.

   56, 82, 84, 88, 91, 92, 95, 95, 97, 100

2. To find the median, you use a method of elimination. What this means is that you eliminate a piece of data on each end of the ordered set, until you reach the middle.

~~56~~, 82, 84, 88, 91, 92, 95, 95, 97, ~~100~~
~~56~~, ~~82~~, 84, 88, 91, 92, 95, 95, ~~97~~, ~~100~~
~~56~~, ~~82~~, ~~84~~, 88, 91, 92, 95, ~~95~~, ~~97~~, ~~100~~
~~56~~, ~~82~~, ~~84~~, ~~88~~, 91, 92, ~~95~~, ~~95~~, ~~97~~, ~~100~~

3. Since there are two numbers remaining in the center of the data, you find the average of the two to determine the median. $\frac{(91+92)}{2} = \frac{183}{2} = 91.5$

The median of this data set is 91.5.

Another way to find the median is by creating a dot plot to organize the data. The same process of elimination works on dot plots, as well. You begin at the extremes and cross them off as you move your way to the center of the plot. The result would be the same.

The median is most useful in describing a set of data when there are extreme values or outliers that would skew the mean. To use the median to describe the data, there should also be no real major gaps in the middle of the data set. The median is also used in constructing box plots, also known as box-and-whisker plots, which will be discussed in the displaying data section of this chapter.

Mode

The *mode* is the numerical value that appears most in a set of numbers. There may be a single mode, more than one mode, or no mode at all. When each value occurs only once in a data set, there is no mode.

It is possible to have a mode of 0 in a data set. Do not confuse a mode of 0 with no mode. These are two different things.

Usually the mode is used for categorical data because it highlights the most common category. However, the mode can be used to describe the data where there are many identical data points. When the mode is used as the measure of center, it describes what is typical about the data set.

The standards at the sixth-grade level tend to encourage students not to use mode as the measure of central tendency. This is because it can provide an inaccurate description of the overall data, if by chance the most common value is far away from the rest of the data, and considered an outlier.

What is an outlier?
An *outlier* is a number in a set of other numbers that doesn't seem to belong. It stands out from the other numbers because it is excessively higher or lower than the majority of the data.

If the mode is used to describe a set of data where the mode is considered an outlier, it can be very misleading to the reader about the overall data set.

Measures of Variation

Measures of variation, also known as measures of spread, are ways to measure how much a collection of data is spread out. A measure of variation uses a single number to describe how the values vary in the set of data. The most common measure of variation is the range, but students in sixth grade also learn about mean absolute deviation.

Range

The *range* of a data set is the difference between the maximum value and the minimum value. It is found by subtracting minimum from the maximum. For example, if the highest test score is a 100% and the lowest test score is a 56%, the range is found by subtracting the two numbers.

$$\begin{aligned}\text{Range} &= 100 - 56 \\ &= 44\end{aligned}$$

The range of the test scores is 44%.

Why is knowing the range beneficial in understanding a set of data?
The range communicates to the reader just how far apart that data is. If the range is small, then you know the data is close together. If the range is large, you know that the data is spread out.

Mean Absolute Deviation

This is a mouthful and sounds a lot harder than it really is. The *mean absolute deviation* is an average of how far each data point in a set is from the overall mean of the set of data. Basically, it tells you how far away the data values are from the mean, or it is the "average distance from the average," as described by Utah District Consortium.

A small mean absolute deviation tells you that most values are close to the mean. A large mean absolute deviation tells you that the data values are more spread out.

The Utah District Consortium has done a nice job of outlining the necessary steps to calculate mean absolute deviation.

1. **Step 1:** Find the mean of the set of data.
2. **Step 2:** Determine the deviation or difference of each number in the data set from the mean.
3. **Step 3:** Find the absolute value of each deviation from Step 2.
4. **Step 4:** Find the average of the absolute deviations from Step 3.

The result of this calculation is your mean absolute deviation.

Using the test result data from a sample math test, the scores are: 95, 56, 82, 97, 95, 84, 92, 88, 91, 100. Let's determine the mean absolute deviation.

1. **Step 1:** $\text{Mean} = \frac{(95+56+82+97+95+84+92+88+91+100)}{10}$

$$= \frac{880}{10}$$

$$= 88$$

The average of the set of data is an 88%.

2. **Step 2 and 3:** By ordering the data from least to greatest, it may help with this process of determining how far away each data point is from the mean. Also, by creating a table, you can organize your data better.

56, 82, 84, 88, 91, 92, 95, 95, 97, 100

Since a table was created, you could incorporate steps 2 and 3 and calculate them together.

Value in Data Set	Deviation of Number from the Mean	Absolute Value of Deviation
56	$56-88=-32$	$\lvert -32\rvert =32$
82	$82-88=-6$	$\lvert -6\rvert =6$
84	$84-88=-4$	$\lvert -4\rvert =4$
88	$88-88=0$	$\lvert 0\rvert =0$
91	$91-88=3$	$\lvert 3\rvert =3$
92	$92-88=4$	$\lvert 4\rvert =4$
95	$95-88=7$	$\lvert 7\rvert =7$
95	$95-88=7$	$\lvert 7\rvert =7$
97	$97-88=9$	$\lvert 9\rvert =9$
100	$100-88=12$	$\lvert 12\rvert =12$

3. **Step 4:** Find the average of the absolute deviations.

Here you can take the numbers from the last column of the table and find the mean.

$$\text{Mean} = \frac{(32+6+4+0+3+4+7+7+9+12)}{10}$$

$$= \frac{84}{10}$$

$$= 8.4$$

8.4 is the mean absolute deviation. By analyzing the information about range and mean absolute deviation, you can draw conclusions about the data. If the range is 44, and the mean absolute deviation is relatively small at 8.4, this tells you the majority of the data is close together.

The mean absolute deviation also tells us a bit more information. Using the previous example, with a mean absolute deviation of 8.4 and a mean of 88, we are able to say that each student scored on average 8.4 percentage points above or below the mean of 88.

Another method for determining mean absolute deviation is to use a dot or line plot. The initial step is still the same: You need to calculate the mean of the data set. Once you have determined the mean, mark it on the line plot and make it stand out with a highlighter. Next, determine the distance that each point is from the mean by counting using the number line. Record these distances. Once you have completed these steps, add up all the distances and divide by the number of data points. The result will be the mean absolute deviation.

Using a Graph

We use data displays to help organize information in a way that allows others to see a visual representation of a numerical set. Data displays also help make it easier to analyze and describe data. It serves as an informative tool.

Graphs are useful in describing data, as they give you a visual picture of how the data are dispersed. Histograms and dot plots can show you the overall shape of the data. Is it curved? Is it straight? Are there patterns? Does it have clusters of numbers? Are there peaks? Are there gaps? Is there any symmetry in the graph? Being able to describe this information also helps to create a broader understanding of the data set as a whole.

A *cluster* of data is a grouping of numbers that are close together in values. A *gap* is indicated by an interval that contains no data. An *outlier* is a number in the data set that is much larger or much smaller than the rest of the set.

Can you imagine reading an article in a magazine or a newspaper loaded with data and no visual to accompany it? That would be horrendous.

Displaying Data

There are three data displays that sixth grade students will focus their attention on. The goal is for students to be able to create and interpret dot plots (a type of line plot), histograms, and box plots (box-and-whisker plots).

Dot Plots

A *dot plot* is a method of graphically displaying data using dots positioned above a number line. A dot plot, although a form of line plot, differs because instead of using x symbols to represent the frequency of a value, a dot is used in its place.

To construct a dot plot you must first begin with a number line that extends from at least the minimum to the maximum. It can go a few numbers beyond the minimum and the maximum, if desired. It can be used to show both categorical data and numerical data. If a dot plot is used to represent numerical data, it is important to remember that a tick mark is required for each consecutive number in the interval pattern. If the data ranges from 5 to 12, but has no data points for 7, you still must create a number line that includes 5, 6, 7, 8, 9, 10, 11, and 12. By omitting any tick marks for values that have no data, you change the visual display of the data set. Check your child's work to make sure she hasn't made this mistake; although seemingly minor, it can cause her to make mistakes in how she describes the data. To represent the data, a dot is placed above the line for each corresponding value. Another important factor is to always title your graph. If a graph does not have a title, the reader has no idea what the data is representing.

The following values represent how many books each of the 20 children in class read over the summer: 2, 3, 5, 1, 7, 4, 2, 0, 2, 3, 2, 2, 2, 3, 0, 1, 5, 3, 3, 2. Let's organize these from least to greatest: 0, 0, 1, 1, 2, 2, 2, 2, 2, 2, 2, 3, 3, 3, 3, 3, 4, 5, 5, 7. Now let's create a dot plot.

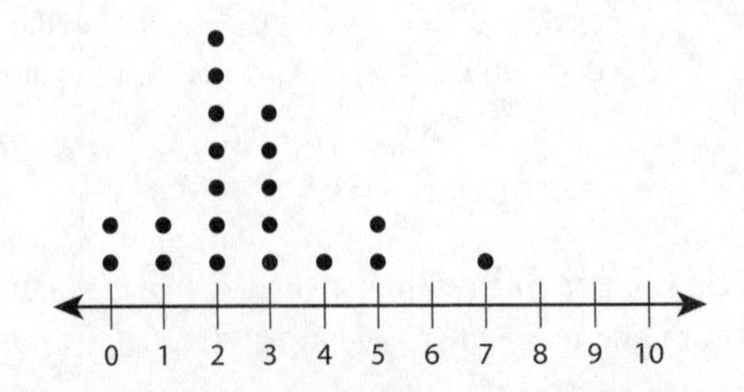

Let's describe the data using measures of center and spread.

The mean of the data is $52 \div 20 = 2.6$

The median is 2

The mode is 2

The range is $7 - 0 = 7$

The mean absolute deviation is 1.26, found by dividing 25.2 (sum of deviations from mean) by 20. The distance each value is from the mean is recorded in the following table.

Value	0	0	1	1	2	2	2	2	2	2	2	3	3	3	3	3	4	5	5	7
Distance from the Mean (2.6)	2.6	2.6	1.6	1.6	.6	.6	.6	.6	.6	.6	.6	.4	.4	.4	.4	.4	1.4	2.4	2.4	4.4

Since the mean absolute deviation is a small number, this tells us the data is close together. In looking at the dot plot, this conclusion is confirmed.

There are advantages of using a dot plot over other data displays. First and foremost, a dot plot is easy to construct, and also easy to read. It preserves all the individual values of the data set, as well. It is best used for small amounts of data, otherwise it becomes cumbersome to construct. Another advantage to a dot plot is that you can easily see gaps and clusters in the data, as well as outliers.

Mistakes can occur when interpreting what exactly the dot means. Some students might "count" each dot as one rather than the value it is representing. For instance, a dot placed over the number seven on the number line represents a value of 7 in the data set.

If your child takes his time in constructing the dot plot so that it is neat and organized, chances are he will do well with this form of data display.

Histograms

A *histogram* is a data display for numerical data in which bars are used at various heights, much like a bar graph, which is used solely for categorical data. There are several differences between a bar graph and a histogram. A histogram uses groups of numbers to create a range for each interval; a bar graph uses category names. A histogram has no spaces between columns unless there is no data for that particular interval; a bar graph typically does. A histogram does not provide the specific data values of the set due to the clustering of the data into an interval; a bar graph tells you the exact number of people that picked that category. Although they have several differences, they do share one thing in common: They both use a y-axis to help readers determine the value at the height of each bar.

There are several advantages to using a histogram to display data. A histogram is often very easy to read. It is also great for representing large amounts of data, because the numerical data can be grouped together into smaller intervals and represented together on a graph.

With every advantage also comes a risk. Students can often make a mistake when breaking the data into equal intervals. If there is not the same number of numerical values in each subset's range (the interval), then the presentation of the data can be skewed. They may also make a mistake in reading the scale on the y-axis, thinking that every block is one when in fact it could be counting by fives, for example. With attention to detail, all of these mistakes and misconceptions can be avoided.

To construct a histogram, begin by making a frequency table to sort the data. To determine the labels in your table as well as the labels for the bars, determine the range of your data and then divide it into equal intervals. Record the data in the appropriate interval by using a tally mark.

A frequency table is a way to summarize data. It tells the reader the number of times a data value occurs. It is created by organizing data values in ascending numerical order, and uses a tally to represent each time the value appears.

Next, use graph paper to draw an x-axis (horizontal line) and a y-axis (vertical line). The y-axis should be drawn at the left end of the x-axis. Be sure to draw the lines long enough so that each box is able to represent an interval of numbers, and that the total range of the data can be displayed. The y-axis typically uses a scale, where one box represents multiples of 1, 2, 5, 10, 20, or 25. Label both axes with a title and the appropriate intervals/scale.

Now that the outer shell has been created, it is time to display the data. Draw a bar above each interval that is the height that correlates with the frequency. Do this for each interval, remembering not to leave spaces between the bars unless there is no data to be displayed. At this point, your child may think she is all done, but not yet. Did she remember to title her data display? Every graph needs to have a title, otherwise we have no idea what the data is talking about.

Let's go through the process of creating our own histogram. Let's say we completed a survey of 50 people, asking them how many movies they have seen at a movie theatre in the last year. Here is the frequency table with all of the data organized.

▼ NUMBER OF MOVIES SEEN IN ONE YEAR

Number of Movie Ranges	Tally	Frequency
0–5	|||	3
6–10	||||	4
11–15	~~||||~~ ~~||||~~	10
16–20	~~||||~~ ||	7
21–25		0
26–30	||	2
31–35	~~||||~~ ~~||||~~ ~~||||~~	15
36–40	||||	4
41–45			1
46–50		0
50–55	||||	4

Using this data we can build a histogram.

Number of Movies Seen in One Year

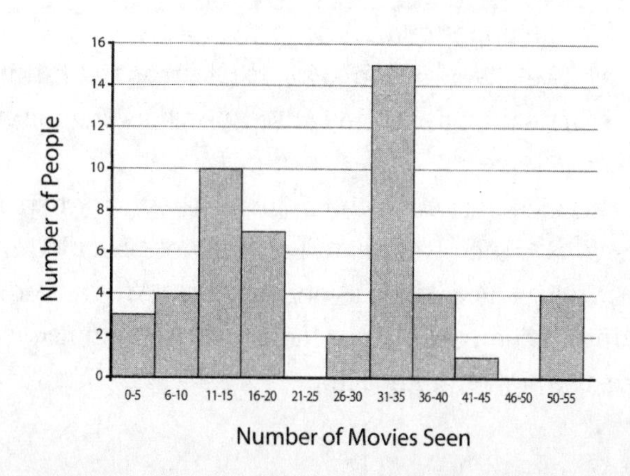

Histograms are a very useful tool for anyone who is in project management or wanting to improve quality of service. Think about wait time. Consumers hate to wait. A study could be done collecting data on how much time customers had to wait to be provided service for a question regarding their cable bill. How many of us have heard "This call is being recorded for quality assurance," "Your wait time is five minutes," or "You are the tenth caller, please hold for the next available representative"? These are pieces of data that businesses use and analyze to determine what they can do to better their level of service. Think about how you could help your child collect data to improve one aspect of this world.

Box Plots

A *box plot*, also known as a *box-and-whisker plot*, is defined by the Common Core State Standards Glossary as "a method of visually displaying a distribution of data values by using the median, quartiles, and extremes of the data set. The box shows the middle 50% of the data, and then extended *whiskers* show the remaining 50% of the data." Wow, that is a lot packed into a definition. Let's break it down into what you know, and what you don't know. You know about median; it is the middle number. When you hear the word "quartiles," what do you think? Are you breaking the data into quarters? Let's look at some of the major elements of a box plot before we get into the nitty-gritty details.

First, you need to know that a box plot is formed from a number line. A box plot summarizes five pieces of data into one nice graph. It provides:

- The median
- The lower quartile
- The upper quartile
- The lower extreme, also known as the minimum
- The upper extreme, also known as the maximum

The nice part about box plots is they can display a lot of data with very little work. Since each data point is not provided in a box plot, this display gives you a general idea about the data. Another advantage is that if you are looking to compare two or more data sets, multiple box plots can be drawn above the same number line.

What Exactly Is a Quartile?

A quartile is exactly what it sounds like: it breaks an ordered set of data into four equal parts. The first quartile, known as the lower quartile or Q1, is the median of the lower half of an ordered data set. The second quartile, known as the median or Q2, is the median of the entire set of data. The third quartile is the upper quartile or Q3, and is the median of the upper half of data.

Remember that the data needs to be in ascending (or increasing) order for this to work. The lowest numbers should be on the left, and the highest number on the right.

Since the quartiles break the data up into four equal-sized parts, each quartile represents 25% of the data; students and adults often misunderstand this. Students also struggle with how to find the lower quartile and upper quartile, but if they know how to find the median of a data set, then this is just finding the median of the first half of the data, and then the median of the second half of the data. It really isn't all that bad.

Another vocabulary phrase that you should become familiar with is the *interquartile range*. This is the difference between the upper quartile (Q3) and the lower quartile (Q1). This value tells us how far apart the majority (the middle 50%) of the data exists.

Once again, the Utah District Consortium has done a great job of creating a set of step-by-step directions that are worth sharing.

1. Write the data in order from least to greatest.
2. Draw a horizontal number line that can show the data in equal intervals.
3. Find the median of the data set and mark it on the number line.
4. Find the median of the upper half of the data. This is called the upper quartile (Q3). Mark it on the number line.
5. Find the median of the lower half of the data. This is called the lower quartile (Q1). Mark it on the number line.
6. Mark the lower extreme (minimum) on the number line.

7. Mark the upper extreme (maximum) on the number line.
8. Draw a box between the lower quartile and the upper quartile. Draw a vertical line through the median to split the box.
9. Draw a "whisker" from the lower quartile to the lower extreme.
10. Draw a "whisker" from the upper quartile to the upper extreme.

You have now created a box plot.

Here is an example of constructing a box-and-whisker plot from scratch. Although this example deviates from the steps outlined here, you can see the same result is attained. First, you need to begin with a set of data.

40 Random Test Scores	
63	80
66	79
78	93
70	90
64	70
75	98
66	92
82	88
81	95
63	89
46	92
59	92
75	90
79	85
85	77
94	59
90	79
83	79
88	99
89	96

40 Test Scores in Numerical Order	
46	82
59	83
59	85
63	85
63	88
64	88
66	89
66	89
70	90
70	90
75	90
75	92
77	92
78	92
79	93
79	94
79	95
79	96
80	98
81	99

Using the ordered list, you can easily determine the median of the whole data set. Since there is no exact middle value, you must add the 81 and the 82 and divide by two. The median is 81.5.

To determine the lower and upper quartiles, you can break the data into two parts:

1. Lower Quartile Data:

46, 59, 59, 63, 63, 64, 66, 66, 70, 70, 75, 75, 77, 78, 79, 79, 79, 79, 80, 81

Again, since there is an even number of data points, we must average the two middle numbers. The lower quartile (Q1) is 72.5.

2. Upper Quartile Data:

82, 83, 85, 85, 88, 88, 89, 89, 90, 90, 90, 92, 92, 92, 93, 94, 95, 96, 98, 99

The median of the upper quartile (Q3) is 90. The interquartile range is figured out by subtracting 72.5 from 90.

$$90 - 72.5 = 17.5$$

This means that 50% of the data lies within a range of 17.5 numbers.

You also need the minimum of 46 and the maximum of 99 to create the box-and-whisker plot. Now that you have the five-point summary completed, let's build the box plot.

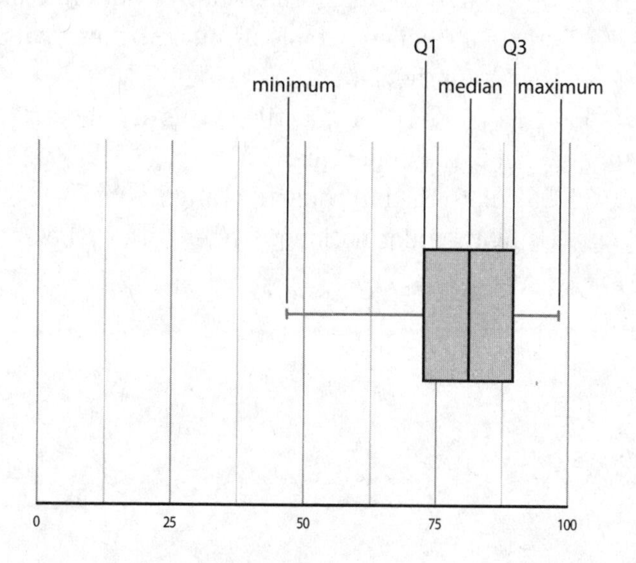

And there you have it. The box represents the central 50% of the data and is a graphic representation of the central trends of one-variable data.

With a little help from YouTube, using the following link, *www.youtube.com/watch?v=ucWmfmXb1kk*, you can use Excel or similar spreadsheet software to double check your calculations. When drawing by hand consider using graph paper as it is gives you a nice neat line with boxes to work from. Happy calculating!

Practice Problems

Using the sample data provided, answer each of the following questions. A survey was completed that asked 25 random people how many years they have lived in their current home. Here is a list of the results.

Data: 5, 2, 7, 24, 53, 18, 13, 5, 2, 11, 14, 25, 36, 2, 6, 8, 15, 28, 30, 1, 15, 18, 21, 25, 16

1. What is the mean?
2. What is the median?
3. What is the mode?
4. What is the range?
5. Determine the mean absolute deviation. Using the data provided, create a frequency table using equal intervals, and then draw a histogram of the data set.
6. What is the lower quartile of this set of data?
7. What is the upper quartile?
8. Calculate the interquartile range.
9. Using the information provided, draw a box-and-whisker plot.

PART 3

7th Grade Common Core Standards

CHAPTER 10

What Is Expected of My 7th Grader?

The Common Core State Standards have stepped up their game and increased the level of difficulty for each grade level, and seventh grade is no exception. What once might have been a general seventh-grade curriculum has been revved up to include many pre-algebra concepts, including graphing proportional relationships on a coordinate plane and exploring its slope. Further studies include scaling two-dimensional models in preparation for learning about congruence and similarity in eighth grade.

What Are the Critical Areas?

As stated by Corestandards.org, students will spend the majority of their year focusing on four critical areas:

1. Developing understanding of and applying proportional relationships
2. Developing understanding of operations with rational numbers and working with expressions and linear equations
3. Solving problems involving scale drawings and informal geometric constructions, and working with two- and three-dimensional shapes to solve problems involving area, surface area, and volume
4. Drawing inferences about populations based on samples

Critical Area One

The goal for students in this critical area is to build upon what they have learned in sixth grade regarding ratios, rates, and proportions. It is in seventh grade that students use this understanding to be able to solve a wide range of percent problems relating to real-world applications such calculating tax, tip, discounts, percent of change (increase or decrease), and interest.

Students also apply their understanding of ratios and proportionality to solve problems about scale drawings. A *scale drawing* is a drawing that has been enlarged or minimized, but maintains the original features of the object. *Proportions* are used to determine similarity between corresponding lengths of objects, as well as understanding that if two objects are similar, meaning they resemble the same shape but are different sizes, then the relationship between side lengths is preserved.

Also with the study of proportions comes a link to graphing on a coordinate plane. Students will learn to graph situations on a coordinate plane to decide if two quantities form a proportional relationship. Ratios are equivalent and form a proportion if the graph is a straight line that passes through the origin. Students then use this graph to explain the meaning of the ordered pair, given the terms of the situation.

Critical Area Two

In this critical area, students learn that all rational numbers can be represented in various ways, including as a fraction, decimal, or a percent. Decimals that are finite or have a repeating decimal are considered to be rational numbers as well. Students continue their work with addition, subtraction, multiplication, and division of all rational numbers, and apply the properties of operations learned in sixth grade to rational numbers, as well.

Students also extend their understanding of integers, the set of positive and negative numbers, to integer operations. Students are expected to investigate, explain, and interpret the rules for adding, subtracting, multiplying, and dividing positive and negative numbers. No longer is it just about memorizing the rules; students are expected to be able to justify why the rules work. Furthermore, all of this information is used to formulate expressions and equations, using at most one variable to solve problems.

Critical Area Three

Geometry is the focus of this critical area. Here students further develop their understanding of area acquired in grade six to include solving problems involving the area and circumference of circles, as well as the surface area of three-dimensional objects. At this point, students are expected to know the formulas for the area and circumference of a circle, and to use them to solve problems. Students also are expected to be able to derive the formula for the area of a circle using the formula for circumference. (This informal way of deriving the formula is really neat, so just wait until Chapter 14.)

Along with exploring area and surface area, students explore what happens when a three-dimensional figure is sliced, much like a loaf of bread or a block of cheese. Knowing what figure is formed by the cross section allows students to determine area, surface area, and volume of two- and three-dimensional objects composed of triangles, quadrilaterals, polygons, cubes, and right prisms.

Students continue their studies with geometry to explore relationships concerning two-dimensional figures. Here they will learn to compute actual lengths and areas from a scale drawing, and will learn to scale figures to different sizes. In practicing this skill, they learn that corresponding side lengths of the scaled figure must be in proportion to the side lengths of the original. Students should be able to work between different scales to complete these tasks. This work is the foundation of learning needed for understanding congruence and similarity in eighth grade.

Seventh graders also learn to construct geometric shapes using a ruler and protractor (and technology, when appropriate) when given particular conditions. For example, given side lengths of 2 cm, 5 cm, and 7 cm, can a triangle be drawn? The answer is, no. The triangle inequality theorem states that in order for a triangle to be constructed, any side of a triangle must be shorter than the sum of the other two sides. If it is longer, then the sides would not be able to meet and a triangle could not be formed.

Lastly, students explore angle relationships when lines intersect. These angle relationships include supplementary, complementary, vertical, and adjacent angles, and this information can be used to solve multistep problems. This understanding helps students to solve for missing angle measurements by writing an equation showing the relationship between the known and unknown values.

Critical Area Four

The last major focus area of seventh grade is that of statistics and probability. Students build upon their knowledge of single data distributions from sixth grade to now be able to compare two data distributions and answer questions relating to differences between these two data sets, or populations. Students begin to explore random sampling to generate data sets, and then make generalizations and inferences based on the results.

Students will also explore probability of a chance event as a number between 0 and 1. Students will learn the differences between theoretical probability (what is expected to happen) and experimental probability (what actually happens).

Students will run experiments to collect data on a single event, and answer questions related to such experiments. Students will learn to develop probability models to collect and organize data, and use such models to determine the likeliness that an event will occur.

Lastly, students will investigate compound events. An example of a compound event is rolling a 2 and then rolling a 4 on a die. Using tools such as organized lists, tables, tree diagrams, and simulations are wonderful ways to determine the probability of a compound event. Students will then work with the probabilities to express them as fractions, decimals, and percents, as a way to communicate the likeliness that the event will occur.

In Summary

After reading about the major focus areas, you will notice that expressions and equations is not a "critical focus area," but do not be alarmed. Students will spend time during their seventh-grade year learning to solve multistep problems with positive and negative rational numbers. They will use these in word problems, two-step equations, $px+q=r$, $p(x+q)=r$, and solving and graphing inequalities in the form $px+q>r$ and $px+q<r$.

Let's put it this way: Seventh-grade students will be busy! Continue reading to learn more details and examples about the standards for mastery in seventh grade.

CHAPTER 11

Ratios and Proportional Relationships

For a concept as straightforward a ratio is, the Common Core's take on it involves some seemingly complex reasoning. That is not to say that any of it is unimportant; oh no, it's all good stuff. Remember, though, it boils down to this: No matter how crazy a problem you are dealing with, it all comes back to the simple idea that a *ratio* is a comparison of two numbers, and a *proportion* is a comparison of equivalent ratios. That's it. Now, that being said, let's peel back some layers of this onion and see what's in store!

What Should My Child Already Be Able to Do?

Students enter seventh grade with an understanding of the meaning of a ratio, and the different ways in which it can be expressed: as a fraction ($\frac{a}{b}$), using a colon (a:b), using the word "to" (a to b), or as part-to-whole relationship (a out of b). They learn that there are different ways to compare two numbers, expressing ratios as part-to-part or as part-to-whole. Students learn to represent these ratios using images, or by taking an image and turning it into a ratio. Along with understanding how to write ratios, students learn to simplify ratios to lowest terms. They do this by using something called a *scale factor*, where they divide a fraction by a fancy form of 1, any number over itself, for example, $\frac{2}{2}$, $\frac{5}{5}$, or $\frac{n}{n}$.

Sixth graders also know about rates and unit rates. They use these to express two quantities with different kinds of units, such as beats per minute, miles traveled per amount of time, etc. The equivalent fractions showing unit rates form a proportional relationship. For example, if a heart beats 300 beats in 5 minutes, this is expressed as the rate $\frac{\text{300 beats}}{\text{5 minutes}}$ and can be simplified to show a unit rate of $\frac{\text{60 beats}}{\text{1 minute}}$, forming the proportional relationship of $\frac{\text{300 beats}}{\text{5 minutes}} = \frac{\text{60 beats}}{\text{1 minute}}$.

At the end of sixth grade, students have learned several ways of applying ratios and rates to solve problems. These methods include using a unit rate, using the scale factor method to simplify or enlarge ratios, using ratio tables, using tape diagrams, and using double number lines.

Sixth graders also extend their understanding of ratios by finding a percent of a number, using methods such as benchmark percents, mental math calculations using 100%, 10%, and 1%, fraction multiplication, decimal multiplication, and double number lines. The various approaches give students several ways to solve a problem, using a method that makes the most sense to them. Students are able to determine values for two types of percent problems: what is the percent of a number, and find the whole when given a percent and a part. There will be a third type of problem that comes into play here in seventh grade.

Lastly, when exploring percent of change and margin of error, students will be using absolute value, another topic that has been learned in sixth grade. *Absolute value* measures the distance a number lies from zero on a number line. Since distance cannot be negative, absolute value must be at least zero.

It is important that your child have a solid understanding of these sixth-grade skills before diving into seventh-grade concepts. Although teachers always integrate some form of review, it is impossible for a seventh-grade teacher to reteach everything learned in sixth grade.

Analyzing Proportional Relationships

Quick, 1 out of 2 is the same as 25 out of what? What did you get? That's right: 50. That wasn't difficult, for a couple different reasons. First of all, $\frac{1}{2}$ is a very common fraction. But more importantly, the proportion $\frac{1}{2}=\frac{25}{x}$ can be solved in one direct step. You probably didn't even think about that proportion when you came up with 50, but you certainly did use it. That proportion can be solved by a simple scale factor of 25; the 1 gets multiplied by 25, and so does the 2.

But what happens when the fractions aren't common, the numbers aren't nice, and the process isn't intuitive? Cross products, that's what happens. Recall that given any proportion, the products of the diagonal pairs will always be equal:

$$\text{If } \frac{a}{b}=\frac{c}{d}, \text{ then } ad=bc$$

Technically, this concept could have been used to solve the opening problem of this section. It just wasn't necessary. We can verify the connection with a little multiplication: $1 \times 50 = 2 \times 25$. Yep, it checks out.

In the proportion $\frac{a}{b}=\frac{c}{d}$, *b* and *c* are called the *means*, while *a* and *d* are the *extremes*. Therefore, another way to describe the concept of cross products would be "the product of the means is equal to the product of the extremes."

Let's suppose a recipe for mashed potatoes serves 4 people and calls for $2\frac{2}{3}$ cups of mashed potato flakes. (Nobody said it was a fancy dish being made.) However, you want to make nine servings instead of four. How many cups of flakes should you use? This problem is obviously a lot trickier than the first one. Let's set up a proportion.

$$\frac{2\frac{2}{3}}{4}=\frac{x}{9}$$

Wait! Don't close the book . . . it's not that bad. Look at proportions as templates. Every proportion is made up of four components. Regardless of the level of complexity of a proportion, it will have top-left, bottom-left, top-right, and bottom-right values. If one (or more) of those values are mixed numbers, negatives, decimals, huge numbers, or variables, so what? The cross products will always be found the same way.

$$\left(2\frac{2}{3}\right)(9)=4x$$

Two times nine is eighteen $(2\times9=18)$ and two-thirds of nine is six $\left(\frac{2}{3}\times9=6\right)$. Eighteen and six make 24, so . . .

$$24=4x$$
$$x=6$$

You need six cups of flakes. Were you expecting a crazy number? Nope. Six. Math is funny like that. Sometimes, the scariest looking of problems have nice, neat answers.

Some teachers will give unique names to the concept or process of using cross products. "Bubble X" refers to the shape made by circling the diagonals of the proportion. "The fish method" gets its name from the design drawn if you follow the path of operations performed within a proportion to get the value of a variable (which looks like a fish).

When solving a proportion, first check to see if you can find the answer directly. In the proportion $\frac{1\frac{1}{4}}{6} = \frac{x}{18}$, the denominator of the first fraction is multiplied by three to get the second denominator. Therefore, just triple the $1\frac{1}{4}$ to get the final answer of $x = 3\frac{3}{4}$. If there is not a direct path (or if you don't see a direct path), go back to the cross products. They work every time.

Determining Proportionality in Relationships

A child's mother partially filled a sippy cup with 2 ounces of juice and 6 ounces of water. Later on, the child's father completely refills the cup with 3 ounces of juice and 10 ounces of water. The child takes a sip and throws the cup at the father.

The father's mixture was not in proportion to the mother's mixture. But how can you tell? By setting up a "would-be" proportion and checking the cross product, you can see if the ratios are proportional. Does $\frac{2}{6} = \frac{3}{10}$? Only if $2 \times 10 = 3 \times 6$ . . . and it does not. So no, the new mixture was different, and the child let the father know about it.

If you wanted to know what fractional amount of the drinks were juice, they are not $\frac{2}{6}$ and $\frac{3}{10}$. Those ratios compare juice to water, or a part-to-part ratio. The first drink had eight total ounces. The second one had thirteen. The mother's mixture was $\frac{2}{8}$ (or 25%) juice. The father's mixture was $\frac{3}{13}$ (or about 23%) juice.

Of course, it is possible for more than two fractions to be compared at once. The following prices were paid for meteorites:

Weight in Grams	Price Paid
4	$480
10	$1,200
6	$800
9	$1,080

Is there a proportional relationship between the weight and the price paid? You could set up a bunch of proportions, each comparing two ratios at a time. Or you could find the unit price of each meteorite, and check to see if they are all the same.

$\frac{\$480}{4} = \$120$ per gram $\qquad \frac{\$1{,}200}{10} = \$120$ per gram

$\frac{\$800}{6} = \$150$ per gram $\qquad \frac{\$1{,}080}{9} = \$120$ per gram

It only takes one "hiccup" to throw the whole thing off. The relationship is not proportional. Every single unit rate would have needed to be equal in order for the relationship to be perfectly proportional.

Ratio and Percent Problems

An entire book could be written on percent problems. The percent is incredibly versatile. It pops up in many different places. Actually, any time two values are being compared, a percent very well could factor into the work being done. In no way whatsoever could every possible percent scenario be covered in this one section. We will, however, explore a few of the most common ones.

The Percent Proportion

Percents are ratios where the denominator is always 100. So any ratio can be converted into a percent by creating a proportion. For example, if we want to know what $\frac{3}{16}$ is as a percent, we can set $\frac{3}{16}$ equal to a fraction over 100 and then solve for x:

$$\frac{3}{16} = \frac{x}{100}$$
$$16x = 300$$
$$x = 18.75$$

This number represents the percent. $\frac{3}{16}$ is 18.75%.

Pretty handy little thing, the percent proportion is. And it can be used for much more than just converting fractions into percents.

For example: What percent of 40 is 24?

The given ratio is 24 out of 40, or $\frac{24}{40}$. So, you write $\frac{24}{40}=\frac{x}{100}$. Using cross products, $40x=2{,}400$ and $x=60$. 24 is 60% of 40.

Okay, maybe that was just another "convert a fraction into a percent" problem. However, the wording was just a little bit different, and that's going to be important in other problems that *are* much more different than simple conversion questions.

It is possible for the "part" to be greater than the "whole." This happens in cases where the percentage is greater than 100%. Therefore, you cannot simply put the smaller number on top of the fraction. Here's a good rule to use: In a word problem, "is" refers to the number that is the part, and "of" refers to the number is the whole. Another way to look at the percent proportion is $\frac{is}{of}=\frac{percentage}{100\%}$.

Let's explore another application of the percent proportion within a word problem. An $800 television set is marked 30% off. How much is the sale price television set? You can be clever and use some number sense first. If it's 30% *off,* then the sale price must be 70% of the original price. Essentially, you want to know what 70% of $800 is.

$$\frac{x}{800}=\frac{70}{100}$$

$$100 \bullet x=56000$$

$$x=560$$

The television now sells for $560.

There is also a percent equation that is sometimes a lot faster than the percent proportion. It requires the percentage to be converted into a decimal: *Part* = *Decimal* × *Whole*. Fill in the two values that you know, and solve for the value that you don't. Some teachers will reveal both methods; some won't. Some teachers will allow students to use either method, and some teachers will expect students to know them both.

One more time with the percent proportion: 36 is 72% of what number? This time, you know the partial amount, and you are looking for the whole amount.

$$\frac{36}{x} = \frac{72}{100}$$
$$72x = 3600$$
$$x = 50$$
$$36 \text{ is } 72\% \text{ of } 50$$

That problem had no context. The Common Core is going to have the students do much more than just crunch numbers, however. Here's a better example where it is going to be necessary to make some sense of what is being asked first.

At the end of the month, a store manager sees that she has taken in $48,686.40. If that amount includes an 8% sales tax, how much money did her store actually bring in?

$48,686.40 represents 100% of the sales plus the 8% sales tax. So that total number is 108% of the total sales. This question is really asking: *48686.40 is 108% of what number?*

$$48686.40 = 1.08x$$
$$48686.40 \div 1.08 = (1.08x) \div 1.08$$
$$45080 = x$$
$$x = 45080$$

The store took in $45,080, before tax. Notice that this was one of the cases where the "part" was greater than the "whole." If this seems confusing, read the final statement out loud and see if it sounds like it makes sense:

"48,686.40 is 108% of 45,080." Yes, it makes sense for the first number to be bigger, since it is more than 100% of the 45,080.

Percent of Change

There really isn't a big difference between percent of change and the percent problems from the previous section on the percent proportion. Percent of change is just a specific example and requires a little bit of numerical sense in the beginning of the process.

Consider this: your spouse comes home, excited, bragging about saving $15 at the store. Sound like phenomenal savings? It all depends, really. Was the purchased item a car? A television? A pair of jeans? The $15 savings needs to be compared to the original price of the item. And so we get the formula for percent of change:

$$\textit{Percent of Change} = \frac{\textit{Amount of change}}{\textit{Original amount}} * 100\%$$

For example, a store buys microwaves for $60 and then sells them for $75. What was the percent of change in the price?

A little bit of simple subtraction tells you that the store tacks $15 onto the price before selling the appliance to you. The $15 is compared to the original $60 (*not* the new $75!) with the fraction $\frac{15}{60}$, which divides to 0.25 and then converts to 25% after multiplying by 100. In this example, the percent of change is called a markup.

When a price increases, the percent of change is called a *markup*. When the price decreases, it is called a *discount*. In percent of change problems that do not deal with money, the words "increase" and "decrease" can be used to describe the percent of change.

A man went on a diet. He dropped down to 180 pounds from his original 200 pounds. His percent of change can be described as a 10% decrease. The 20 pounds he lost is compared to the original 200. $\frac{20}{200} = 10\%$.

Margin of Error

Margin of error is a way to express how far away an estimate or a guess is from an actual (or correct) amount using the ratio of the *amount* of error as compared to the actual amount. This margin of error is almost always expressed as a percent. These problems really can be lumped into the same section as percent of change. The process is so similar that you may not even be able to tell the difference between the two types of problems.

$$Margin\ of\ Error = \frac{|Experimental - Actual|}{Actual} * 100\%$$

The "experimental" could refer to a guess, an attempt, an estimation, or a number obtained from an actual experiment.

The vertical bars around the numerator of the fraction represent absolute value. It means that you are not concerned with the sign of the number, just the number itself. In margin of error problems, it is okay to be off either below or above the actual amount.

A carnival has a contest where people guess how many jellybeans are in a large jar. You guess 450, and your friend guesses 550. It turns out that there were 500 jellybeans in the jar. Your margin of error is $\frac{50}{500}$, or 10%. But so is your friend's. Two different guesses can tie with the same margin of error.

It is possible for a margin of error to be greater than 100%. If somebody had guessed 1,200 jellybeans when there are actually 500, they would be off by more than the actual amount. This would create an improper fraction ($\frac{700}{500}$) and ultimately a percent greater than 100% (140% margin of error).

Again, it would be practically impossible to give an example of every single type of percent problem that your student may see in middle school. The important skills will be being able to extrapolate the known partial amount, whole amount, and/or percentage from the problem, as well as being able to use either the percent proportion or the percent equation to find the missing value.

Practice Problems

Solve for x in the following proportions:

1. A certain recipe that serves 5 people calls for $1\frac{3}{4}$ cups of flour. How many cups of flour will you need if you want to serve 8 people?
2. A recipe that serves 10 people calls for $1\frac{1}{2}$ cups of sugar. How many servings would you be able to prepare if you had 21 cups of sugar?

For questions #3 and #4, decide whether or not the ratios are proportional.

3. $\frac{6}{7}$ and $\frac{7}{8}$
4. $\frac{1\frac{1}{3}}{4}$ and $\frac{2\frac{1}{2}}{7\frac{1}{2}}$
5. Convert $\frac{13}{40}$ into a percent, using the percent proportion.
6. A television is purchased by a store for $400. It is then marked up 30% before it goes on the sales floor. After a while of not being sold, it is marked as 25% off. At what price is the television selling now?
7. An $80 pair of jeans is now being sold for $64. Find the percent of change for the discount.
8. A food bill came to the table with the 8% meal tax already included. 15% of that amount was then added as a tip. If $37.26 was left on the table, what was the cost of the meal *before* tax?
9. A 160-pound woman wants to drop down to 140 pounds. What is her target percent of change?
10. Scott guessed that a $400 exercise bike was $330. Tina guessed that a $16,000 car was $13,500. Find each of their margins of error to determine who made the better estimate.

CHAPTER 12

The Number System

In this chapter, we will venture left of zero on the number line. We will open our minds to the numbers that, until now, were all but ignored—the negative numbers. Negative numbers are going to be sprinkled throughout every single type of operation and almost every single type of application from here on out, so it is safe to say that they are kind of a big deal. Cheer up, it's time to get negative!

What Should My Child Already Be Able to Do?

In sixth grade, students started to explore the number set of integers and learned how these numbers represent quantities both greater than and less than zero, extending in opposite directions on a number line. They learned situations that would require the use of both positive and negative values. For instance, 10 feet below sea level could be expressed using the integer −10 ft. A temperature of 45°F would be expressed using a positive integer of 45°F. They also learned about absolute value, $|n|$, which measures the distance a number is from 0 on the number line. For instance, both 5 and −5 have the same absolute value. They are both a distance of 5 away from 0 on the number line, just in opposite directions. Understanding the basic idea of what an integer is and what it represents is the foundation for learning about integer operations, and eventually operations of rational numbers.

Also at this point in the game, students should know how to add, subtract, multiply, and divide whole numbers, fractions, and decimals using the traditional algorithms for each, and complete these operations with fluency and ease, an important skill for integer operations.

Students in sixth grade focused primarily on the operation of division. Their studies included division of whole numbers and division of fractions. Mastery of the sixth-grade skill of dividing using long division will be important as seventh graders learn to convert a rational number to a decimal. Sixth graders also learned how to model division of fractions by fractions. It was important for them to conceptually understand what it means to divide a fraction by a fraction, and then progress to learning about the algorithm to divide fractions: To divide fractions, multiply by the reciprocal of the second fraction. The next step for students is to compute with positive and negative fractions, which are encompassed in the set of rational numbers.

Fraction Operations

There was a time, not all that long ago, when teachers devoted a huge percentage of middle-school classes to meticulously teach kids the method of adding, subtracting, multiplying, and dividing fractions. Now the expectation is that students learn these processes much more quickly (starting as early as elementary school!) so that more intricate work involving fractions can be done. It will be a huge advantage for students to know their fraction arithmetic at the beginning of the year, so that they are not forced to begin this important unit with a bunch of grueling review.

This is a time when a lot of people develop their fear of fractions. None of the procedures within fraction arithmetic are really that difficult; it's not multi-variable calculus. But just like anything else, if pushed too hard, too fast, an anxiety will be created. Before you know it, the "I hate fractions, they're too hard!" comments will fill your home. So just in case your student never picked up on the fine art of fraction arithmetic, or even if he simply forgot some of the details, let's take a close look at each of the four major operations, stripped down to each of their basic procedures.

Adding and Subtracting Fractions

It's fairly clichéd when working with fractions, but imagine a pizza. Cutting the pizza into slices means each slice is a fraction of the entire pie. The denominator of a fraction determines the size of the slice. Notice that as the denominator gets bigger, the slice (or the fraction) gets smaller.

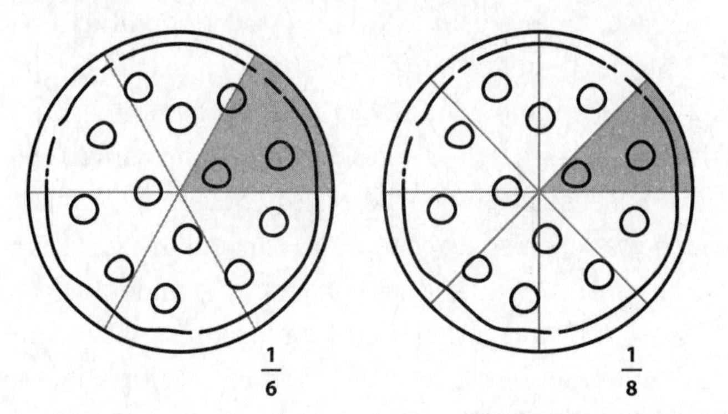

If you took one slice of each pizza, it would be unfair to consider them as equal slices. You certainly can't add them together and call them two of either size—they're not two-sixths; they're not two-eighths. In order to get each pizza divided into equal-sized slices, you need to do some more cutting. If each slice of the first pizza is cut in half, then there will be twelve slices. If they were cut into thirds, then there would be eighteen slices. The possibilities for amounts of slices are the multiples of six: 6, 12, 18, 24, 30 . . . . Similarly, the possibilities for amounts of slices in the second pie would be the multiples of eight: 8, 16, 24, 32 . . . .

Twenty-four is the first number to appear in both lists. This is an important number, called the *least common multiple*, or LCM. Take a look at the same amounts of pizza as earlier, only now with equal-sized slices:

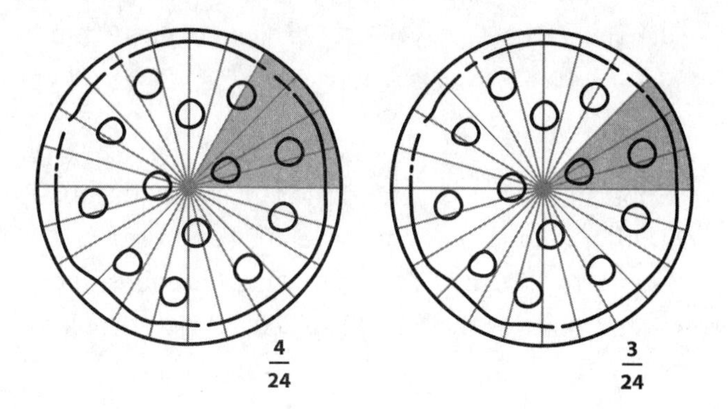

You can now combine the same amount of pizza and all of the slices will be the same size, enabling us to add them together. $\frac{4}{24}+\frac{3}{24}=\frac{7}{24}$.

No, a twenty-fourth of a pizza is not very practical in real life, but it did model the problem well.

When adding or subtracting fractions, no matter the size, complexity, or sign of those fractions, you must first convert them so that they share a common denominator. Although any common multiple will work, the least common multiple will minimize the amount of work necessary.

Here's another example, straight to the point:

$$\frac{3}{4}+\frac{5}{6}=$$

$$\frac{9}{12}+\frac{10}{12}=\frac{19}{12}=1\frac{7}{12}$$

Notice that this sum resulted in an improper fraction. It was then converted into a mixed number. This conversion is not necessary, but it is commonly expected since $1\frac{7}{12}$ is easier to conceptualize than $\frac{19}{12}$.

It is a very common mistake to add the denominators as well as the numerators. Don't. Remember that the denominator only refers to the size of the "slice." It is the numerator that refers to how many of that slice you have. The denominator acts very much like a unit of measurement.

Subtraction follows the same rules as addition. The fractions must have a common denominator in order to be subtracted. Sometimes, it becomes necessary to "borrow" from the whole number in order to perform the subtraction. Consider the following subtraction problem:

$$4\frac{3}{5} - 2\frac{5}{6} =$$

Look specifically at the fractions. You are asked to take $\frac{5}{6}$ away from $\frac{3}{5}$.

The problem is, $\frac{5}{6}$ is bigger than $\frac{3}{5}$ ($\frac{25}{30}$ as compared to $\frac{18}{30}$). You need to borrow from the whole part of the mixed number. Continuing with the pizza analogy, this is very much like opening another box of pizza and cutting it up into slices.

$$4\frac{3}{5} - 2\frac{5}{6} =$$

$$4\frac{18}{30} - 2\frac{25}{30} =$$

$$3\frac{48}{30} - 2\frac{25}{30} = 1\frac{13}{30}$$

By opening up one of the four pizzas, you get 30 more slices to add to the eighteen that were already sitting there. You are then able to subtract the 25 from the 48.

Multiplying Fractions

Great news! The toughest part is over. Multiplying fractions is the easiest of the four operations. Simply multiply the numerators and then multiply the denominators. That's it. The only thing that may get in your way is if the fraction is part of a mixed number. If so, convert it to an improper fraction.

How do I convert a mixed number into an improper fraction?
If we once again use the pizza analogy, a mixed number is like having some complete pizzas in closed boxes and a few slices in an open box. Converting a mixed fraction into an improper fraction is like opening *every single box* up and presenting *all* of the pizza as slices. Mathematically, this means to multiply the whole integer by the denominator and add the numerator. This value becomes the new numerator. The denominator stays exactly the same as it was.

Here's an example of a fraction multiplication problem that involves converting from a mixed number to an improper fraction:

$$\left(\frac{4}{9}\right)\left(3\frac{3}{8}\right)=$$

$$\left(\frac{4}{9}\right)\left(\frac{27}{8}\right)= \text{(The 27 comes from } 3\times8+3.\text{)}$$

$$\frac{4(27)}{9(8)}=$$

$$\frac{108}{72}=\frac{3}{2}=1\frac{1}{2}$$

It should be noted that at the point of $\left(\frac{4}{9}\right)\left(\frac{27}{8}\right)$, some canceling out could have been done to make the problem a little easier. We can simplify the 4 and the 8 each by 4, since one ends up in the numerator and the other in the denominator. Similarly, 9 can be simplified out of the 27 and the 9 for the same reason. Now the problem becomes $\left(\frac{1}{1}\right)\left(\frac{3}{2}\right)$, which is obviously $\frac{3}{2}$ or $1\frac{1}{2}$.

Dividing Fractions

If you can multiply fractions, you can divide fractions. Just one little tweak to the division problem can turn it into a multiplication problem. Dividing is the same as multiplying by the reciprocal. A reciprocal is when the fraction in perfect rational form (that just means a fraction with no integer in front) gets flipped upside down. For example, $\frac{2}{3}$ and $\frac{3}{2}$ are reciprocals. 4 and $\frac{1}{4}$ are reciprocals as well.

Make sure that you only flip the second fraction—the fraction you are dividing *by*. Leave the first fraction alone.

For example, let's divide 16 by $\frac{8}{9}$.

$$16 \div \frac{8}{9} =$$

$$16\left(\frac{9}{8}\right) =$$

$$\frac{16}{1}\left(\frac{9}{8}\right) =$$

$$\frac{2}{1}\left(\frac{9}{1}\right) = \frac{18}{1} = 18$$

Since a division can be represented by a fraction, then division of fractions can be expressed as a fraction made of fractions. This is called a *compound fraction*. It can look a little intimidating, but just dissect it into a division problem with the more comfortable ÷ symbol, and you will be fine.

$$\frac{\frac{2}{3}}{\frac{3}{8}} =$$

$$\frac{2}{3} \div \frac{3}{8} =$$

$$\frac{2}{3}\left(\frac{8}{3}\right) = \frac{16}{9} = 1\frac{7}{9}$$

Integer Operations and Absolute Value

If you wanted to be really, *really* technical, then –4 would not be pronounced "negative four." It would be pronounced "the opposite of four." To help visualize this, consider a number line:

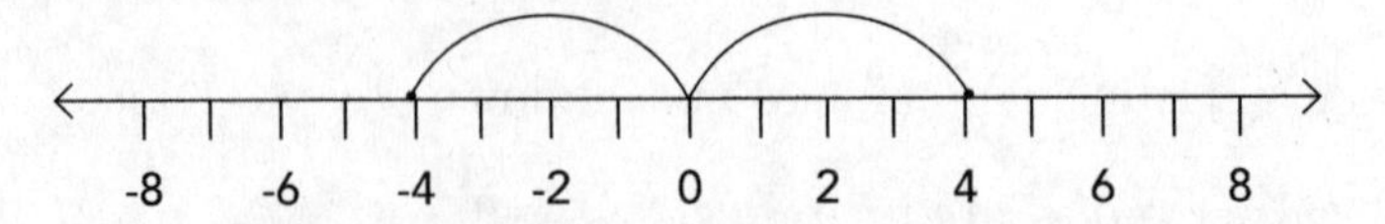

Negative four is the same distance away from zero as positive four. The "–" symbol lets you know to travel to the left side of the zero point on the number line, but to a point the same distance from the zero. Therefore, the opposite of –16 becomes 16.

Absolute Value

The distance between a given value and zero on the number line is called the *absolute value*. It is denoted by vertical bars around the quantity. For example:

$$|-5| = 5$$
$$|0| = 0$$
$$|13| = 13$$

Since absolute value is a distance, the smallest possible result is zero. That being said, if absolute value is combined with other operations the final result may be negative, but something would have to happen after the absolute value function.

$$7|3-8|-40 =$$
$$7|-5|-40 =$$
$$7(5)-40 =$$
$$35-40 =$$
$$-5$$

The vertical bars of absolute value (called rods) are treated like parentheses. If an operation appears inside them, that operation is performed first. When the absolute value of that result is found, the rods turn into parentheses, and then order of operations continues as far as possible.

Integer Operations

There are countless models that are used in the classroom to help students add, subtract, multiply, and divide positive and negative numbers. The most common model is the integer chip. It is usually red on one side and yellow on the other. Teachers declare one color as "positive" and the other as "negative." The first concept introduced with these chips is the *zero pair*. That is to say that if you have one positive and one negative, they combine together to "cancel out," leaving you with zero.

It wasn't until the Mayan civilization came up with it that people considered zero as a quantity worth mentioning. Previously, numbers were always used to count *something*. If you had nothing, it was not important. This was obviously a huge step forward in the world of math as we know it today.

Adding Integers

It comes as no shock that if two positive numbers are added together, the result is also positive: $7+4=11$ because four of an item combined with seven of an item makes eleven of that item. Very similarly, adding two *negatives* together results in a negative. The negative part is like a label or a unit of measurement. Three dollars plus five dollars is eight dollars. Three cheeseburgers plus five cheeseburgers is eight cheeseburgers. So three negatives plus five negatives equals eight negatives:

$$-3+(-5)=-8$$

The parentheses around the -5 in the previous problem do not serve as a separate operation. They are only there to lessen the chance of confusion. $-3+(-5)$ is the same thing as $-3+-5$, but some teachers consider the latter of those equations "bad form."

Things get a little bit more interesting when adding numbers with different signs. This is where the zero pair comes into play. The problem $5+(-7)$ can be modeled with integer chips as follows:

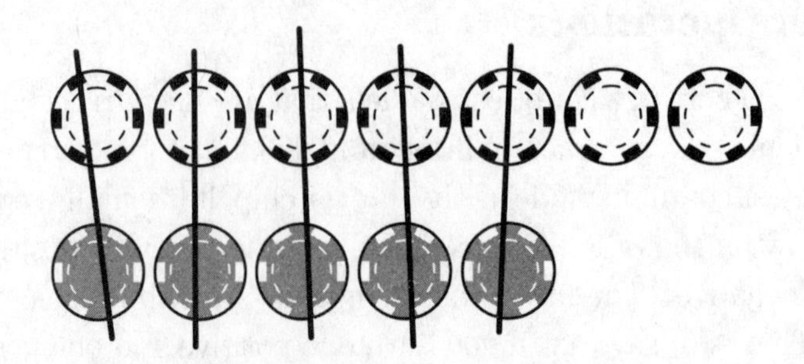

When combined altogether, the collection will have five zero pairs that will cancel out. All that will be left are the two negative chips. Therefore, $5+(-7)=-2$.

Well, that is fine and dandy if the numbers in the problem have small absolute values. What happens when we need to add large positive and negative numbers? Who's paying for all of those chips? There must be an easier way.

Here is the easier way. When adding two numbers of opposite signs, the answer will have the same sign as the addend with the larger absolute value. The number attached to that sign will be the difference between the two addends' absolute values.

Hmmm, that seemed a little wordy. Time for an analogy. Picture a sporting event with two teams, the Positives and the Negatives. The result of any addition of opposite signs will be the same as answering the question, "Who won, and by how much?"

For example, let's take a look at $45+(-37)$. If this was a final score, the Positives won forty-five to thirty-seven. It could be reported that the Positives won by eight. From that, we get $45+(-37)=8$.

Subtracting Integers

There are so many different possible scenarios when subtracting positive and negative numbers that it is impossible to create a cute all-in-one rule to tell students what to do. The models that can be used to show students what is happening often do more harm than good by confusing them even more. Therefore, the easiest path to success is going to be to convert any confusing-looking subtraction problem into an addition problem.

How do I convert subtraction into addition?
Subtracting means to add the opposite. The first value is left alone, the subtraction is changed to addition, and the second value is replaced with its opposite. A nifty name for this concept is "Leave, change, change."

The problem $-24-(-68)$ looks awful. This would be a great time to convert into addition. The -24 gets let alone, the subtraction gets turned into addition, and the -68 becomes 68. Now the problem has become $-24+68$.

Using the sporting analogy from earlier, the positives win by 44. Therefore, $-24-(-68)=44$.

Nobody is saying that you *need* to use this trick to subtract. If a problem is easy like $65-45$, then just subtract them and get 20. How difficult a problem needs to be in order to switch it to an addition problem is entirely up to you.

Multiplying and Dividing Integers

These two operations are lumped together because they share the exact same rule. These are very quick rules that work for any multiplication and/or division problem:

> When multiplying or dividing two numbers with the same sign, the result will be positive.
>
> When multiplying or dividing two numbers with different signs, the result will be negative.

The absolute values of the numbers in the problem play no role in the sign of a multiplication or a division problem; only the amount of negative signs do. Another way of thinking about this rule is recalling that the "–" symbol attached to a number means "the opposite of." If the signs in a multiplication problem are both negative, then it would mean that the opposite of the opposite is being taken, rendering the answer positive.

This rule can be extended to a large string of factors in a multiplication problem:

$$-2(-4)(3)(-1)(-5)(4) =$$

There is an even amount (4) of negative factors in this problem. Therefore, the answer is going to be positive. The "opposite of an opposite" happens twice. Instead of trying to keep track of the signs along with doing the multiplication, the sign can be figured out and then the multiplication can be done separately.

Many students make the mistake of using the quick multiplication and division rules while adding and subtracting. Adding and subtracting a long string of numbers must be done one number at a time. You cannot find the sign of the answer by counting how many of the numbers are negative, as you can in multiplication.

When dividing numbers, you may get a nice integer as an answer as in $-\frac{20}{4}=-5$. However, the division may not work out nicely. If this happens, the answer can be left in simplified fractional form. If the result is negative, the sign should either be placed on the numerator or out in front of the fraction. For example, 42 divided by −50 certainly does not work out nicely. It can be represented by the fraction 42/−50, simplified to 21/−25, and then expressed as $\frac{-21}{50}$ or $-\frac{21}{50}$ as a final answer. $\frac{21}{-50}$ *does* mean the same thing, but leaving the negative on the denominator is a form not generally used. Think of the negative sign as a fishing bobber: It's either going to sit on the surface or be up in the air, but never stay beneath.

The other option would be to perform the division by hand using long division or simply with a calculator, and then express the answer as a decimal.

Fraction Arithmetic with Positives and Negatives

Traditional classrooms have been setting students up for culture shock once they venture off into the big, scary real world. Lessons are often taught with one specific learning target; when completed, a new focus is presented with distinct articulation and compartmentalization. But in the real world, problems are often multistep, multifaceted situations requiring a combination of mathematics in order to be resolved.

That being said, it would seem to be a good idea to blend together the last topics into one multistep problem. Consider the following:

$$\frac{4\frac{2}{3}+\frac{5}{6}}{\frac{-22}{3}}\left(\frac{\frac{-7}{3}}{\frac{-10}{3}-\frac{1}{6}}\right)=$$

Wow! A lot of people are going to take one look at that problem and die a little inside. Take a close look at it, though—it's just addition, subtraction, multiplication, and division. Take one step at a time, follow the order of operations, and you'll be surprised at just how "doable" this problem is.

Start with the addition in the first numerator and the subtraction in the second denominator. This will leave you with a multiplication problem of two compound fractions:

$$\frac{4\frac{2}{3}+\frac{5}{6}}{\frac{-22}{3}}\left(\frac{\frac{-7}{3}}{\frac{-10}{3}-\frac{1}{6}}\right)=\frac{4\frac{4}{6}+\frac{5}{6}}{\frac{-22}{3}}\left(\frac{\frac{-7}{3}}{\frac{-20}{6}-\frac{1}{6}}\right)=\frac{4\frac{9}{6}}{\frac{-22}{3}}\left(\frac{\frac{-7}{3}}{\frac{-20}{6}+\frac{-1}{6}}\right)=\frac{5\frac{1}{2}}{\frac{-22}{3}}\left(\frac{\frac{-7}{3}}{\frac{-21}{6}}\right)$$

Now it's time to do some fraction division. Don't let the negative signs scare you—the first fraction consists of a positive divided by a negative. Its quotient is negative. The second fraction consists of a negative divided by a negative. Its quotient is positive. Now that you know the signs of the results, the focus can be put on the fraction arithmetic. The signs can be ignored for now as long as you don't forget to include them wherever appropriate in the final answers:

$$\frac{5\frac{1}{2}}{\frac{22}{3}} \textit{ means } 5\frac{1}{2} \text{ divided by } \frac{22}{3}$$

$$\frac{11}{2}\left(\frac{3}{22}\right)=\frac{33}{44}=\frac{3}{4} \text{ but recall that this quotient is negative} \ldots -\frac{3}{4}$$

$$\frac{\frac{7}{3}}{\frac{21}{6}} \textit{ means } \frac{7}{3} \text{ divided by } \frac{21}{6}$$

$$\frac{7}{3}\left(\frac{6}{21}\right)=\frac{42}{63}=\frac{2}{3}$$

You now have the two factors of the multiplication problem simplified. All that is left to do is the multiplication:

$$\frac{-3}{4}\left(\frac{2}{3}\right)=\frac{-6}{12}=\frac{-1}{2}$$

Can you believe it? After all of that, $\dfrac{4\frac{2}{3}+\frac{5}{6}}{\frac{-22}{3}}\left(\dfrac{\frac{-7}{3}}{\frac{-10}{3}-\frac{1}{6}}\right)$ just ended up equaling $-\frac{1}{2}$!

Converting Rational Numbers to Decimals

Any fraction made up of two integers is going to do one of two things as a decimal. It is either going to *terminate* (meaning the digits end), or it is going to *repeat* (meaning that at least one of the digits of the decimal part recurs infinitely).

It is well known that $\frac{1}{2}=0.5$. But to verify this, you would need to divide 1 by 2. Using the old-fashioned division still taught in elementary school, you get:

$$\begin{array}{r} .5 \\ 2\overline{)1.0} \\ \underline{10} \\ 0 \end{array}$$

Once the 2 divides into the 10 "nicely," the division is done. The decimal terminates.

It may take a few digits before a decimal terminates. Such is the case in $\frac{13}{16}$:

$$\begin{array}{r} .8125 \\ 16\overline{)13.0000} \\ \underline{128} \\ 20 \\ \underline{16} \\ 40 \\ \underline{32} \\ 80 \end{array}$$

It may have taken four digits, but eventually the decimal terminated. Be careful, though—not all decimals terminate. Given the example of converting $\frac{2}{3}$ into a decimal, something new happens:

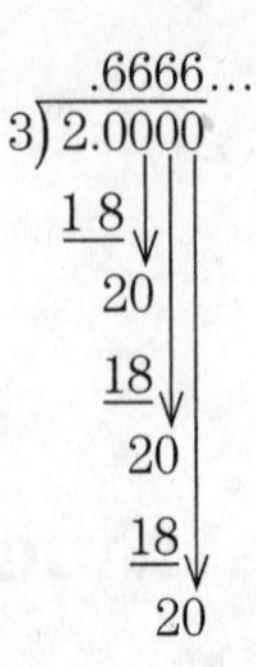

It becomes apparent that an infinite number of 0s will drop down to create an infinite number of 20s, resulting in an infinite number of 6s in the decimal. The 6 is the repeating digit in the decimal form of $\frac{2}{3}$. This is expressed as $0.\overline{6}$.

> The line above a repeating decimal part is called the *vinculum*. Knowing this is not the slightest bit important to a student's success in middle school; it is just a fun fact.

Sometimes, the repeating pattern in a decimal has more than one digit. Consider the conversion of $\frac{3}{11}$ into a decimal:

```
       .2727...
   11)3.0000
      2 2
        80
        77
         30
         22
          80
```

In this example, the underworking of the division toggles back and forth between 30 and 80. This alternately produces 2s and 7s in the decimal. This pattern continues infinitely and so $\frac{3}{11} = 0.\overline{27}$.

Again, every single fraction will either terminate or repeat in decimal form, as long as that fraction is made up of two integers. It would be an entirely different story given $\frac{\pi}{\sqrt{2}}$, since neither the numerator nor the denominator is an integer. Any such example, as well as any decimal that neither terminates nor repeats, is an irrational number and will be explored in more detail in the eighth grade.

Practice Problems

For questions 1–10, simplify completely by performing the operation(s).

1. $|-5|$
2. $|13|$
3. $|6-11|$
4. $-3|4-1|-4$
5. $16+(-27)$
6. $4-(-9)$
7. $4(-9)$
8. $\frac{-32}{-4}$
9. $5-\frac{-10}{2}+(-7)(-6)$
10. $\frac{\frac{3}{4}}{\frac{-2}{3}}\left(\frac{\frac{-2}{5}}{\frac{3}{4}-\frac{1}{2}}\right)$
11. Use long division to determine whether $\frac{8}{11}$ repeats or terminates as a decimal.
12. Use long division to determine whether $\frac{9}{80}$ repeats or terminates as a decimal.

CHAPTER 13

Expressions and Equations

Although math classes in the seventh grade are not officially called pre-algebra, a lot of the work that the students will be doing is very reminiscent of the curriculum in the old-fashioned pre-algebra course from junior high. Students will enter the seventh grade knowing what a variable is, but now is when they will get their hands dirty with many of the properties and uses of variables. But don't worry—there is still plenty of arithmetic as well.

What Should My Child Already Be Able to Do?

In the sixth grade, students first learn exactly what a variable is and how it's used. They will have been exposed to the concept of the variable much earlier.

$$5+?=9$$

Missing addend models like the one here are no different than solving one-step equations other than the fact that a symbol (such as a question mark, star, or smiley face) is used instead of a letter.

Students should enter the seventh grade being able to solve one-step equations. The next logical step will be to solve *two*-step equations in the seventh grade. The equations will not stray from the four arithmetic operations (addition, subtraction, multiplication, and division). The students' prior knowledge of inverse operations will be used frequently.

An inverse operation is one that *undoes* the effect of another operation. Addition and subtraction are inverse operations to one another, as are multiplication and division. For example, if $4 is added to $5, and then the $4 is subtracted from that total, the result is the original $5. Similarly, if $20 gets doubled to $40, dividing by 2 results in the original $20.

Students with deficiencies in reading comprehension may struggle when it comes time to translate between English phrases and sentences and the mathematical expressions and equations they represent. Students will be expected to know keywords that signify each of the four basic operations. These keywords include, but are not limited to:

▼ KEYWORDS IN ARITHMETIC

Addition	Subtraction	Multiplication	Division
Sum	Difference	Product	Quotient
Plus	Minus	Times	"Goes into"
And	Less	Of	Grouped
Total	"Take away"	By	Over

The phrases "less than" and "is less than" sound very similar but mean completely different things. "8 *less than* 10" denotes subtraction, resulting in an answer of 2, whereas the phrase "*is less than*" refers to inequalities. "8 *is less than* 10" is nothing more than a true statement. It should also be noted that "less than" reverses the order of the numbers in a subtraction problem, while "less" keeps the numbers in the same order as they were read. For example, "15 less than 20 is" translates to $20 - 15$ and results in 5, while "15 less 20 is" translates to $15 - 20$ and results in -5.

As is the way often seen in math, there may only be one answer to a problem but dozens of ways to arrive at that answer. Well, in the same way, $6+4=10$ has dozens of ways that it could be written out in words. If this gets confusing, the best solution is to communicate with your child about the math she is doing. Have her speak out loud to describe what is happening within the problem. If you hear the same word being used over and over again, ask her if she knows a different word that means the same thing.

Equivalent Expressions

One of the most important things to remember, if not *the* most important thing to remember, when working with expressions is that they are a *not complete number sentence*. If an equal sign is present, along with an equivalent expression to the right side of that equal sign, then it is an equation, not an expression. Sometimes you will see an equal sign at the end of an expression. This occurs when an expression is being simplified and an equivalent expression is written on the next line. Do not confuse this for an equation.

▼ EXPRESSION VERSUS EQUATION

Examples of expressions:	$3x+4$	7	x^2y	$5\times2-3x+6=$
Examples of equations:	$3x+4=7$	$5\times2-3x+6=3\times2+5x-14$		

Another (albeit obscure) way to think about expressions is with the name Elizabeth. By saying "Elizabeth," no information is given about her. "Elizabeth" is not a complete sentence. We can give her nicknames like Liz, Beth, Lizzy, and Eliza, but we still don't know anything about her. The exact same thing is happening when finding equivalent algebraic expressions. We are giving new names to the expression—it doesn't change the thing, it doesn't give any more information—it is just a new name.

An *algebraic* expression is an expression that has at least one variable in it.

Now consider "Liz is in the seventh grade." This is a complete sentence. Everything following the word "is" tells you something about Liz. In math, "is" translates to "equals." Therefore, equations are full sentences. Equations can be solved to find the value of the variables. But if a variable is embedded within an expression, the value of the variable will not be known. The only action you can take with an expression is to either simplify it (like calling Elizabeth "Liz") or expand it (like calling her "Eeeeelizzzzzzabeeeeeeeeth").

Like Terms

Terms are separated in expressions with + or with −. *Like terms* refers to the variable parts of the terms.

$$3xy^2$$

In this term, xy^2 is called the variable part. It is the part of the term consisting of all of the variables as well as any exponents attached to the variable. The '3' is called the coefficient. It lets you know how many of the variable part are present. The coefficient and the variable part are always connected by multiplication.

If the variable parts of two or more terms are exactly the same, then they are considered like terms and can be combined by adding or subtracting the coefficients. If there is *any* difference in the variable parts, the terms cannot be combined with addition or subtraction.

$5x + 3x =$

$8x$

$5x^2 + 3x =$

$5x^2 + 3x$ (It cannot be simplified any further because the variable parts of the two terms are different.)

$4x^2 + 5x - x^2 =$

$3x^2 + 5x$ (Two of the terms could be combined, but not all three. Notice that since the last term was negative, the terms were combined using subtraction.)

The Distributive Property

Consider the multiplication problem $5(x+4)$. There are two values being multiplied together: 5 and the quantity $(x+4)$. In order to see what the product should equal, you can use a rectangular array:

	x	$+4$
5	$5x$	$+20$

The 5 gets distributed to the x as well as to the 4 by means of multiplication. The distributive property is a lot like handing out the same thing to everyone in the group. Each term inside the parentheses is going to be multiplied by the value on the outside.

Here are some more examples:

$$3(w-4) = 3w - 12$$
$$-6(x+2) = -6x - 12$$
$$y(y-10) = y^2 - 10y$$
$$-z(z-7) = -z^2 + 7z$$

When the value being distributed is negative, make sure that every term inside the parentheses gets multiplied by the negative. It is an extremely common mistake to forget to distribute the negative to the second term.

Factoring Algebraic Expressions

A factor is a value in a multiplication problem. To *factor* something means to break it into a multiplication problem. Although there are a few different types of factoring out there, in this instance we are going to use the distributive property in reverse.

Instead of multiplying everything in parentheses by a value, now the task is to find the value that "goes into" every term in the expression.

Consider $10x - 5$. There are two terms. You may notice that each term is divisible by five. The five gets factored out and is placed in front of parentheses. What is left of each term remains inside the parentheses.

$$10x - 5 = 5(2x - 1)$$

You can always check your work with factoring by distributing the outside value to check to see if the result is the original expression. $10x - 5$ factors into $5(2x - 1)$ because distributing $5(2x - 1)$ results in $10x - 5$.

Technically, one could factor out 1 and end up with $1(10x - 5)$. However, that wouldn't be too earth-shattering; anything multiplied by one equals itself. The instructions for factoring will almost always say to "factor completely." Therefore, you want to look for the *largest* amount that can be cleanly divided into all of the terms.

Add, Subtract, Factor, and Expand Linear Expressions

Now that the ground rules of manipulating algebraic expressions have been established, it is possible to find multiple expressions equivalent to one given. Much like 9 can be rewritten as $8+1$, $7+2$, $6+3$, $5+4$, $3(3)$, etc., there are many ways to rewrite algebraic expressions.

For example, find three expressions equivalent to $12x + 6$.

- Using factoring: $12x + 6 = 6(2x + 1)$

- Using the idea of like terms: $12x+6=10x+2x+1+5$
- Using both concepts: $12x+6=2(5x+8)+2x-10$

Obviously, these are not the only three possible answers. Quite frankly, the possibilities are endless depending on how elaborate one chooses to be.

For a more practical use of the preceding concept, look no further than the following word problem.

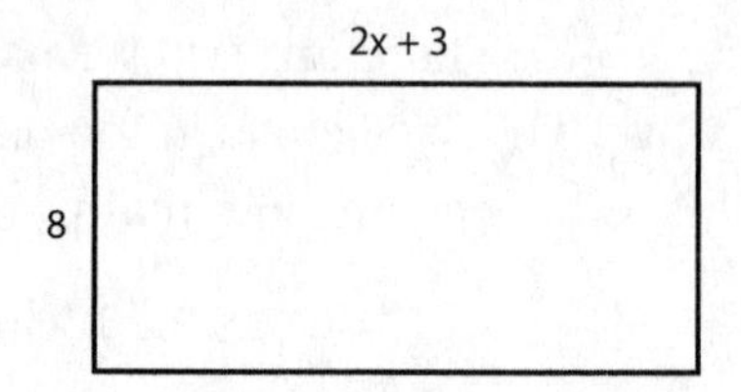

WHICH OF THE FOLLOWING IS NOT AN ACCURATE EXPRESSION FOR THE PERIMETER OF THE RECTANGLE?

1. $8+(2x+3)+8+(2x+3)$
2. $4x+22$
3. $8(2x+3)$
4. $2[(2x+3)+8]$

The third expression would be used to find the area of the rectangle, not the perimeter. All of the other three expressions accurately measure the perimeter.

Using Algebraic Expressions and Equations to Solve Real-Life Problems

Every student can buy into the idea of math's importance in the world of personal finance. Whether money is coming in or going out, there are many factors to how the flow of cash is figured. Often there are formulas that can be used to calculate costs and payments. And, as before, there may be more than one correct way to represent these formulas with an expression.

Percents find their way into finance problems often. It seems strange that when people see an item on sale, they calculate how much is being discounted from the price and then subtract that from the original cost. For example:

There are many ways to come up with the sale price. One way would be to use the traditional approach of finding a percent—multiplying the original cost by the decimal form of the percent:

$$\$40 \times 0.3 =$$
$$\$12 \text{ off}$$
$$\$40 - \$12 \text{ means the shirt is now } \$28.$$

But why not just find the new price? If 30% is being saved, then 70% is being spent:

$$\$40 \times 0.7 =$$
$$\$28 \text{ for the shirt. Done.}$$

Furthermore, a little bit of number sense can be used to make the math even easier. Finding 10% of something is easy; $10\% = \frac{1}{10}$. Divide the price by ten by moving the decimal point one place value to the left. 10% of $40 is $4. So 70% must be seven times $4 . . . or $28.

Many people find 15% a very standard tip for dining out. 15% can be broken into 10% + 5%. Finding 10% is easy and then 5% is just half of 10%. By finding those two amounts separately and then adding them together, 15% can be calculated without need for a calculator.

Coefficients, Variables, and Constants

Johnny is doing some yard work for the sweet lady down the street. He gets paid $2 for every bag of leaves he fills. At the end of the job, if Johnny works well, he'll get an extra $10 as a bonus.

The expression that can be written to represent how much he makes is written in terms of the number of bags he fills. That is to say, that the variable present in the expression represents the number of bags. Choosing b to represent the variable makes perfect sense.

It makes absolutely no difference as to which variable is chosen to represent the unknown in an expression or equation. The answer will not change if x was chosen instead of b. However, choosing the letter that stands for what is being represented is a good idea. It helps recall what the solution to the equation actually means once you finish—a solid first step toward communicating a final answer thoroughly.

The expression $2b + 10$ can be used as a formula for how much Johnny gets paid. Notice that the 2 is written as a coefficient, attached to the variable. The two-dollar amount is directly tied into the number of bags he fills. Each additional bag yields two additional dollars. However, the ten-dollar amount is written at the end, unattached to the variable. Regardless of how many bags he fills, his bonus is ten dollars. This value is called a *constant*.

The neighbor can use this formula to figure out how much to pay Johnny at the end of the job. If Johnny wants to use the formula to figure out how many bags he'd need to fill in order to make a particular amount of money, he's going to need to solve an equation.

Solving Two-Step Equations

In order to solve a two-step equation, two inverse operations must be used to isolate the variable. In addition to deciding which operations to use, it is important to choose the correct order of executing those operations.

Solving Two-Step Equations versus Birthday Presents

Isolating a variable in a two-step equation is a lot like unwrapping a birthday present. When someone gives you a gift, rarely is the item just handed to you as-is. Often the gift is placed into a box and then wrapped up prettily.

Now that the gift is given to you, it's time to find out what you got! Consider the process of opening the gift: You unwrap the paper first and then take the item out of the box.

The two actions taken by the gift giver are being undone in reverse order by the gift receiver. Such is the way with two-step equations. The person who writes the equation performs two operations onto a variable. To solve it, those operations must be "undone" in reverse order.

For example, if Johnny wants to know how many bags he must fill in order to earn \$40, he must solve the equation $2b+10=40$. Focusing on the expression with the variable and using order of operations, it becomes apparent that the variable is first being multiplied by two and then added to ten. To solve the equation, the ten must be subtracted first, and then the two must be divided away. Obviously, those actions must also be done to the right side of the equation in order to keep things equally balanced.

$$2b+10=40$$
$$2b=30$$
$$b=15$$

Johnny must fill 15 bags in order to make \$40.

Although it happens often, it is not always true that the constant is eliminated from the equation first. It is important to consider order of operations. Parentheses can change things up. For example, given $4(x-3)=12$, the variable is being subtracted by three and *then* multiplied by four. Therefore, the first step in solving this equation would be dividing both sides by 4.

As a general rule, word problems should have word answers. Be sure to include a label on the answer or the answer as a complete sentence. If the equation is not part of a word problem, represent your answer as an equation, declaring the variable. For example, write $x=5$, not just 5.

What Is a Solution to an Inequality?

An inequality takes into consideration the other two possibilities when two quantities are being compared—the first can be greater than the second, or the first can be less than the second.

The trichotomy property states that given any two numbers, *a* and *b*, *exactly one* of the following three things must be true: Either *a* is greater than *b* ($a > b$), *a* is less than *b* ($a < b$), or they have the same value ($a = b$).

The symbols ≤ (less than or equal to) and ≥ (greater than or equal to) just allow for either of two different things to be true . . . but at the same time. (A number cannot be both equal to *and* greater than another number!) There is one more symbol that allows for two out of the three possibilities: ≠ allows for the first value to be greater than or less than the second value, but not equal to.

If, instead of an equation, an inequality is given to be solved, the process isolating the variable will be exactly the same as it was in solving equations, but there will be an infinite amount of correct final answers. For example:

Solve the following inequality.

$5x + 12 < 27$

$5x < 15$

$x < 3$

This means that any number less than three will satisfy the inequality.

There is one new thing to keep in mind when solving inequalities. Whenever multiplying or dividing *by* a negative number, the direction of the inequality symbol must be spun around. For example, the solution to $-10x < 30$ is $x > -3$.

There will also be word problems that will require the students to write their own inequalities. Be on the lookout for some common phrases to help translate the English into numbers and symbols.

"Greater than or equal to," "is at least," and "no fewer than" all mean $\geq$.

"Less than or equal to," "no more than," and "no greater than" all mean $\leq$.

"Strictly greater than," and "is more than" both mean $>$.

"Strictly less than," and "is less than" both mean $<$.

Graphing Solution Sets

Since there is an infinite amount of correct solutions to an inequality, the entire set of solutions gets plotted on a number line to serve as a visual representation. It would be impossible to list out every correct answer!

If the inequality symbol is one with a line underneath, it means that the variable *can* equal the number on the other side (called the critical point). In these cases, that value will be plotted on the number line with a filled-in dot. Then shading occurs to the appropriate side of the critical point, dependent upon the direction of the inequality symbol—to the left for less than, to the right for greater than.

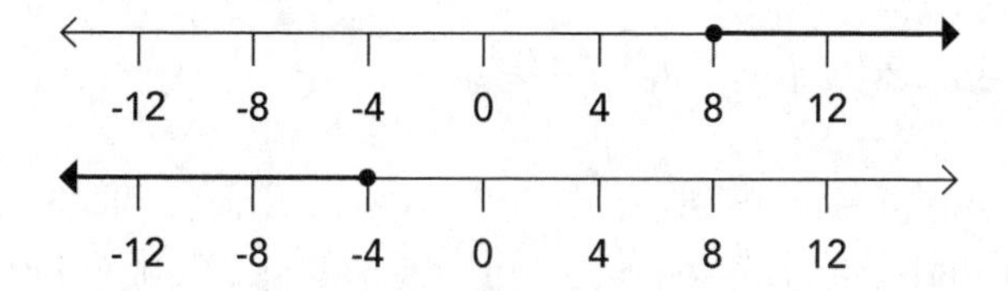

The sign on the critical point has absolutely no effect on the direction of the shading or the type of circle plotted. It just means that the critical point will be on the left side of the zero on the number line.

If the inequality symbol is one *without* a line underneath, then the critical point is plotted with an open dot, but the shading occurs the same as before.

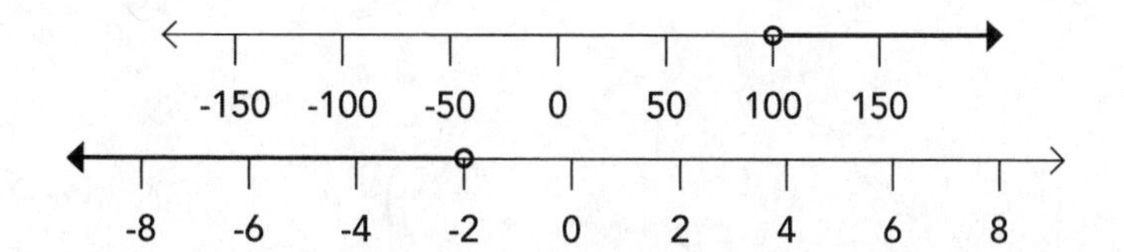

Practice Problems

Label each as an expression or an equation.

1. $3x+7$
2. $x^2+5=9$
3. $2x^4$
4. $3x-4=2x+8$
5. $15=2x$
6. $4x^2y-6$
7. $a^2+b^2=c^2$

Simplify each completely.

8. $3x+2x$
9. $2x^2+5x$
10. $4x+3y-2x+5y$
11. $3x^2+2x^2+10x-7x-6+10$
12. $5(x+4)$
13. $x(x-6)$
14. $-3(y-5)$
15. $2(3z+4)-4z$

A kennel pays Susie to walk dogs at a rate of $5 per dog. She has to go through a $4 toll to go to work as well as to come home.

16. Write an expression that represents the amount of money she makes in terms of dogs walked. Use d for the number of dogs walked.
17. How many dogs did she walk on a day when she made $62?

18. Rephrase this statement as an inequality: Four more than three times a number is no more than nineteen.
19. Solve the inequality from the previous question.
20. Graph the solution on a number line.

CHAPTER 14

Geometry

The best way to make sense of a difficult real-world problem is to first sketch a picture of it. However, without a solid reasoning of the connection between geometry and algebra, that picture is rendered useless. Throughout middle school, students will be exposed to not only area and volume formulas, but also higher-level thinking questions involving geometric objects. The goal here is to get students thinking independently, so that they will be able to reason their way through any geometric problem they see in the future.

What Should My Child Already Be Able to Do?

In a perfect world, all numerical values would work out to a nice neat integer—no decimals, no fractions. Obviously, this is not the case. Students are going to be expected to be able to not only measure lengths accurate to fractions of units, but also calculate things such as area and volume involving non-integer values.

Recall that an *integer* is any number, positive or negative, that has no fractional or decimal parts. Examples include —4, 7, and 0. *Non-integers* are any numbers (again negative or positive) that do have a fractional or decimal part.

Students explore the basic idea of working with non-integer lengths in the sixth grade. Quite a bit of exploration regarding the "how?" and the "why?" behind working with fractional amounts occurs before entering the seventh grade. Therefore, it will be very important for students to be proficient with calculations involving both fractions and decimals if they are going to be successful with seventh-grade geometry.

Students will enter the seventh grade knowing how to use two-dimensional shapes to "cover" a three-dimensional object. Calculating the sum of these areas (called the surface area) will be explored further in the seventh grade. Being able to reason as to which shapes are necessary in order to calculate surface area will be a prerequisite.

Scale Drawings of Geometric Figures

Let's be honest—most of the mathematical real-world problems people face come from situations much bigger than a single sheet of paper. No example illustrates this better than an architect representing the layout of a house with blueprints. Every single line drawn must be in perfect proportion with the corresponding actual lengths of the potential home. The ratio of drawing length compared to the actual length is called the *scale factor*. A scale factor of $\frac{1}{8}$ inch:1 foot means that for every eighth of an inch in the drawing, one full foot of actual length is being represented.

On a much larger scale (quite literally) are maps. Even though impossible-to-refold paper maps are growing more and more scarce, online and smartphone maps still need to abide by the rules of scaling the actual distances proportionally to the image you see on your screen. For the map of a town, an inch on the computer screen may represent one mile (1 inch:1 mile). For a state map, that same inch on the screen may now represent fifty miles (1 inch:50 miles). Of course, those two examples are completely arbitrary—a scale can be set to any desired amount.

Scale drawings are not only used to shrink large things down. They can also be used to scale a very small thing up. However, all of the same rules of proportionality would still be upheld.

In order to calculate an actual distance from a scale drawing, you need to know two things: the scale of the drawing, and the length of the segment. For example:

Find the length of the longest edge of the park.

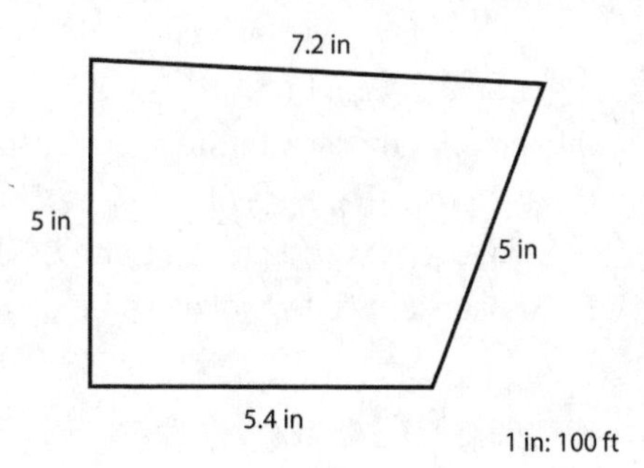

Since the scale is 1 inch:100 feet, the ratio $\frac{1}{100}$ can be used to find any missing length. The longest side is labeled as 7.2 inches long. Although this particular example has a scale that's fairly easy to work with, let's set up the proportion to answer the problem.

$$\frac{1}{100} = \frac{7.2}{x}$$

Multiplying 7.2 by 100 gives us the final answer of 720 feet.

In the preceding example, finding the final answer was a matter of simply multiplying. Sometimes, the scale will be more obscure, such as 3 inches:125 feet. In cases like these, using cross products will be necessary to solve the proportion. For example, given that same drawing with the new scale:

$$\frac{3}{125} = \frac{7.2}{x}$$
$$3x = (7.2)(125)$$
$$3x = 900$$
$$x = 300$$

The actual length becomes 300 feet.

Two- and Three-Dimensional Figures

It would be incomplete and irresponsible to leave the "3-D" part out of the title for this section. However, in comparison, there is little work done in the seventh grade with three-dimensional solids. The overwhelming majority of the work will focus on two-dimensional shapes.

Many people incorrectly classify three-dimensional objects such as cubes and spheres as *shapes*. They are not shapes. Shapes are flat and two-dimensional, like squares and circles. Cubes, spheres, and other three-dimensional objects are called *solids*.

In the seventh grade, the mention of three-dimensional solids comes into play with how some polygons can be constructed from pyramids and prisms. Although an interesting exercise in the connection between shapes and solids, this is not the most important material in this unit. For example, in the following illustrations, both a rectangle and a trapezoid have been created by passing a flat plane through a prism and a pyramid. (Imagine the prism and pyramid are made out of butter and the plane is a large, warm butcher knife.)

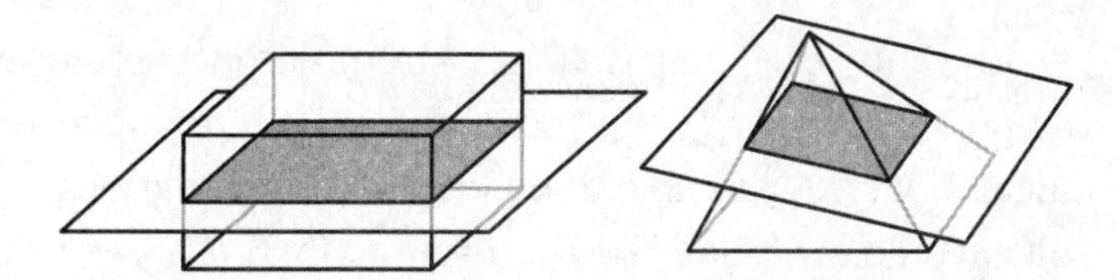

A rectangle sliced from a prism, and a trapezoid sliced from a pyramid.

Many other shapes can be created by passing planes through a three-dimensional solid but again, it is not the most critical piece of geometry in the seventh grade. There will be a much larger focus on classifying polygons, as well as using formulas to find areas and other measurements of traditional shapes.

Area and Circumference of Circles

It is always an important thing for students to be able to see exactly why something like an area formula in math works. It would be easy for a teacher to list off a bunch of formulas and just say, "Here . . . use these." But obviously, that wouldn't be teaching—that would be telling. Big difference.

The Number Pi

All circles are similar. That is to say, that the only thing that differentiates one circle from another are their sizes. This is convenient, because it means that all circles behave the same way as one another. The first connection was found literally thousands of years ago by the Greek mathematician, Archimedes. He discovered that the circumference (the distance around a circle) is proportional to the diameter (the distance across the circle through its center).

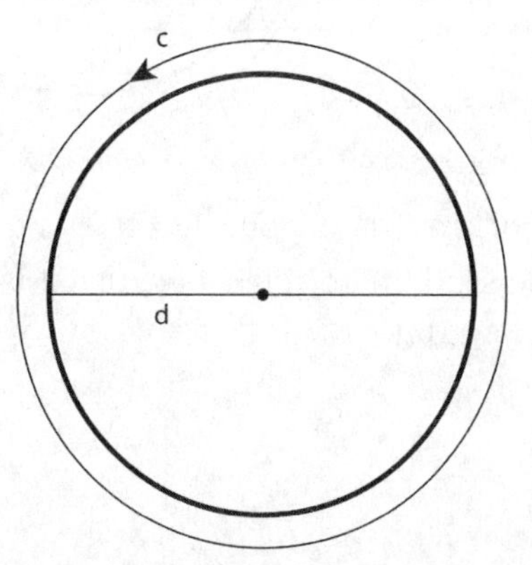

Archimedes found that the circumference is always a little bit more than three times the length of the diameter. Using a series of polygons and calculations, he was able to hone in on that value as accurately as 3.14. This value would eventually be represented by the Greek letter, π.

π, or pi, is an irrational number. That means that it is a decimal that never repeats, nor does it end. Supercomputers have been programmed to continue with the early work of Archimedes, evaluating pi to trillions of decimal places. Although fascinated by the vastness of the number as a decimal, mathematicians often settle on 3.14 as an appropriate approximation for pi.

Circumference

Since it literally is the definition of pi and not a derived formula, it is not with shame that teachers simply have students memorize the circumference of a circle as $C = 2\pi r$. (The diameter in this representation is $2r$, since the diameter of a circle is the same thing as twice the radius.)

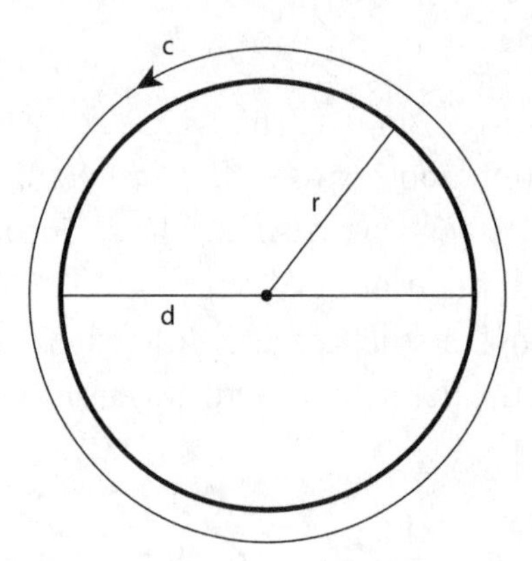

It is important that circumference is one of the first measurements discussed when studying circles, because it is used to explain where other formulas were created.

Area of Circles

Before we discuss the area formula for circles, let's review the area formulas for rectangles and parallelograms. The area formula for a rectangle is easy: Multiply the two dimensions together. It is the method in finding out how many unit squares it takes to cover the shape. For example:

4

3

1	2	3	4
5	6	7	8
9	10	11	(12)

The area was easy to find because of the right angles. However, only a slight tweak needs to be made to see why the area formula for a parallelogram is exactly the same as the formula for rectangles.

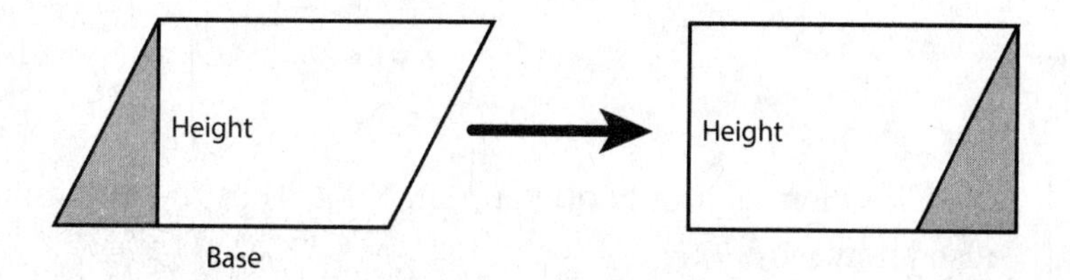

Any parallelogram can be turned into a rectangle with a little cut-and-pasting.

To explain the area formula for circles, let's slice the circle up into a bunch of pieces, like a pie or cake.

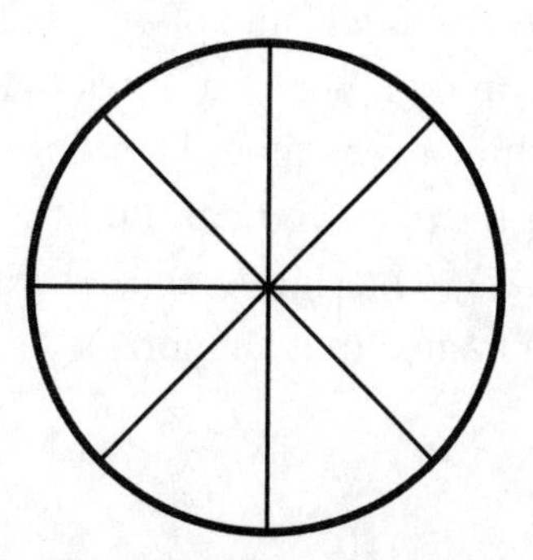

Now, without adding or subtracting from the area, rearrange the pieces like so:

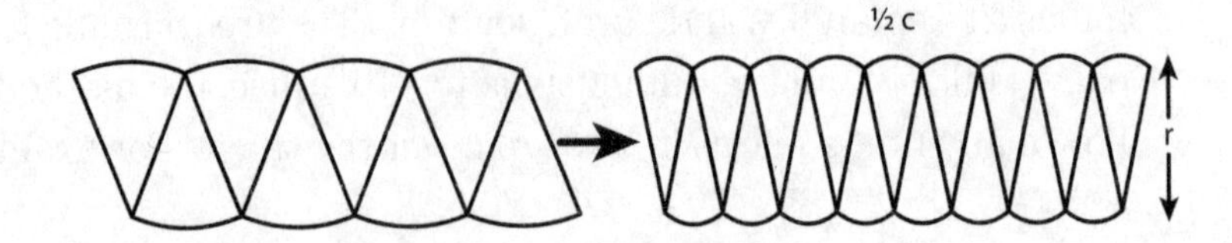

The result looks surprisingly like a parallelogram, doesn't it? As a matter of fact, the more slices, the more it would look like a parallelogram. Making more and more slices, working toward an impossible infinite amount of slices, would, in fact, render this shape into a perfect parallelogram.

The height of the shape is equal to the radius of the circle. The length is equal to half of the circumference. So, the area of the circle must be the radius multiplied by half of the circumference.

$$r \bullet \frac{1}{2}(2\pi r) =$$
$$r \bullet (\pi r) =$$
$$r \bullet \pi \bullet r =$$
$$\pi \bullet r \bullet r =$$
$$\pi r^2$$

Therefore, not only do you now know that the area of a circle $= \pi r^2$, you also know why.

Angles and Angle Relationships

There are ways to show it, ways to prove it, and students will explore them, but simply put, the angle sum of any triangle is 180 degrees. It's true! If you measure the three angles of any triangle and add them together, you will get the same thing every time. Therefore, if you know two of the three angle measures in a triangle, you can find the third by calculating how much of the 180 is missing. The significance of 180 is great enough that a vocabulary term has been named in its honor.

Supplementary Angles

Angles whose angle measures add up to 180 degrees are called *supplementary angles.* Often mentioned in pairs, supplementary angles can be drawn together to form a flat line, called a *straight angle.* (Think of the expression "doing a 180-degree turn.") The supplement to an angle is the amount that has to be added in order to complete the straight angle.

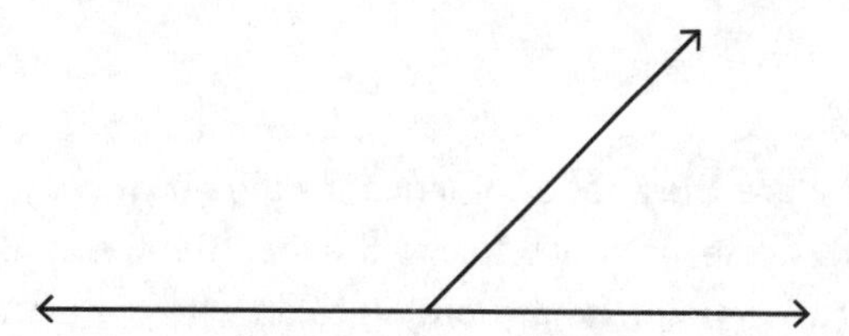

Combining the concepts of triangle angle sums and supplementary angles, you get this neat little trick:

Find the missing value.

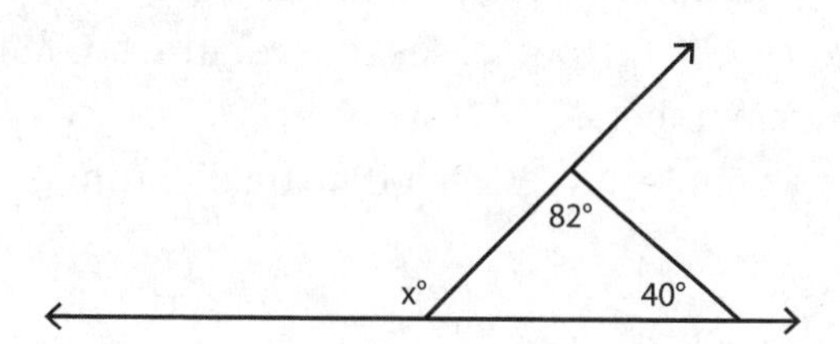

The missing value is the supplement to the unknown angle of the triangle. However, *that* angle must equal the amount needed to complete the 180-degree angle sum. Therefore, the missing angle must simply be the sum of the two given angles—in this case, 122 degrees.

Vertical Angles

When two straight lines intersect, two pairs of vertical angles are created. Vertical angles appear opposite one another within an intersection. The angle measures of vertical angles will always be equal. (Imagine a pair of scissors. As your fingers get further apart, the scissors open. Close your fingers, close the scissors.)

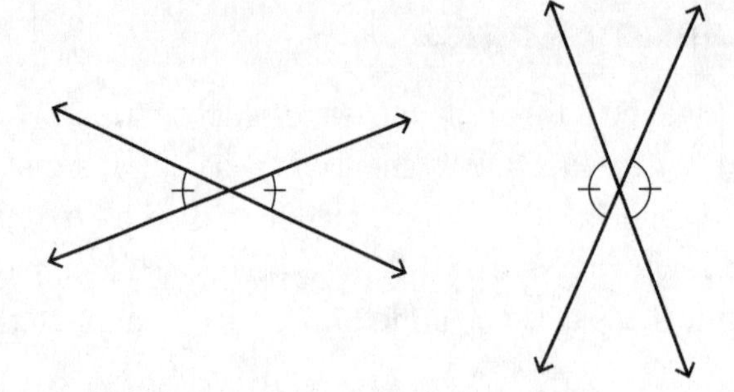

Why are they called vertical angles if no vertical line is drawn?
In this case, vertical refers to the fact that the angles share the same *vertex*, or corner point. It has nothing to do with the direction in which they are drawn.

At this point, students know enough about angle relationships to find missing measurements in some fairly elaborate diagrams. These problems are often multistep processes that require students to think logically as well as crunch numbers.

For example, given the following diagram:

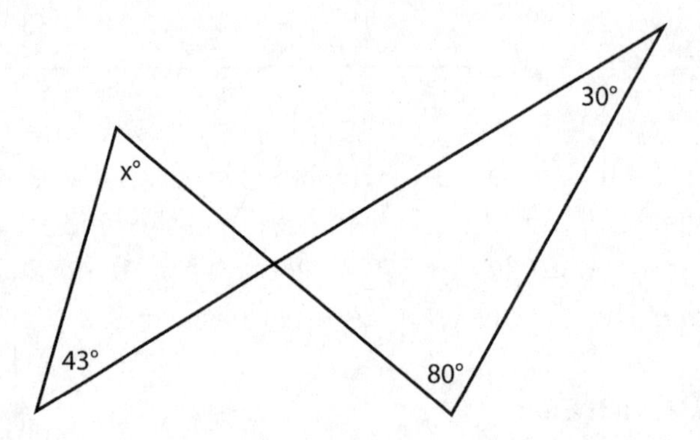

The value of x can be found by utilizing angle sums and vertical angle properties. Can you figure it out? Here's a hint: if you know two angles of a triangle, fill in the third so that the angle sum is 180. Also, fill in any unknown value that makes a vertical pair with a known amount; it must be the same.

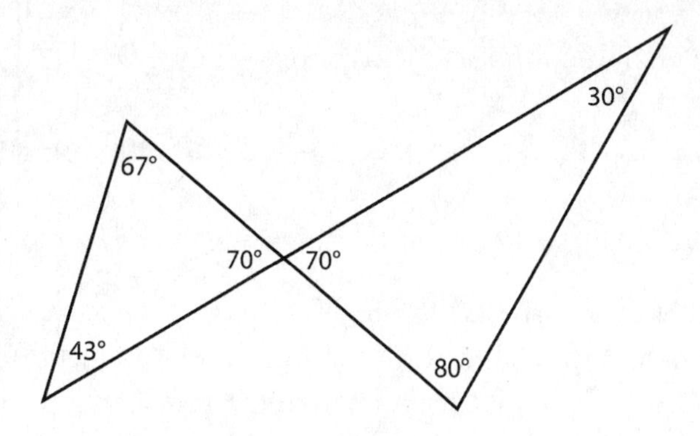

Eventually, you can work your way over to the final answer of $x = 67$.

Area, Volume, and Surface Area of Two- and Three-Dimensional Objects

It is simply not enough for a student to be able to input information into an area or volume formula and find an answer. Students are going to be asked to work backward with the formulas, to reason out what it would take to find a particular measurement within the shape or solid. A very basic example would be as follows:

> What is the height of a triangle whose area is 48 square feet and whose base is 12 feet long?

As always, a good strategy to start this problem off would be to make a sketch of what is known.

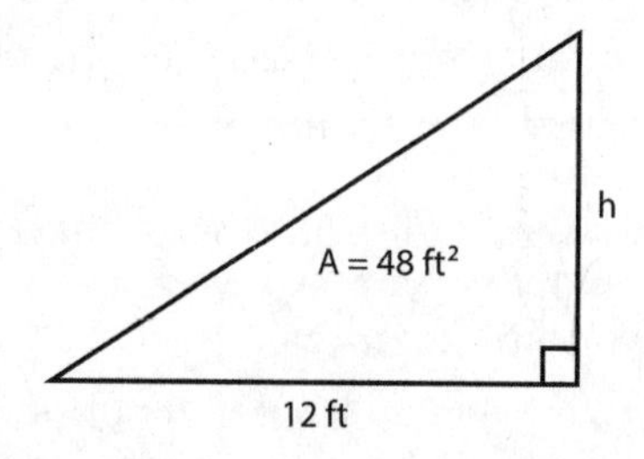

Find the missing height of the triangle.

The area formula for triangles is $A = \frac{1}{2}bh$. The base (which we know is 12) must still be multiplied by $\frac{1}{2}$ and h, and the equation has to yield 48. Since we know $12 \times 4 = 48$, $\frac{1}{2}$ times h must be worth 4. This means $h = 8$. The height of the triangle is 8 feet.

This approach was rather algebraic. It is certainly not the only way that this problem could have been answered. It is up to the student to reason with the problem and decide which method works best for him.

A *heuristic* (pronounced "yur-ISS-tic") is a problem-solving strategy. Making a drawing is one of the most common ones. Some other useful heuristics include (but are not limited to) working backward, making an organized table, and looking for a pattern.

Volume and Surface Area

Seventh graders will only work with volumes and surface area of solids comprised of polygon faces. In other words, nothing with a circle—no cylinders, no cones. There will be questions asked here where students will need to reason backward from a formula in order to get a specific measurement of the solid, very similar to the area problem we just looked at.

However, perhaps more intriguing will be the questions that have students jump from working in one dimension to another. That may sound like a plot from a bad science fiction movie, but it's actually quite straightforward and concrete. For example:

What is the volume of a cube that has a total surface area of 150 cm^2?

Perhaps you are keen enough to notice that there are two completely different types of measurement mentioned in this problem. Volume is a three-dimensional measure, and surface area is the sum of individual areas, which are two-dimensional measures. The common bond, however is that they are both calculated from a three-dimensional object. Oddly enough, in order to answer this problem, you need to travel down to the *first* dimension for a moment, and find the *length* of one edge of the cube.

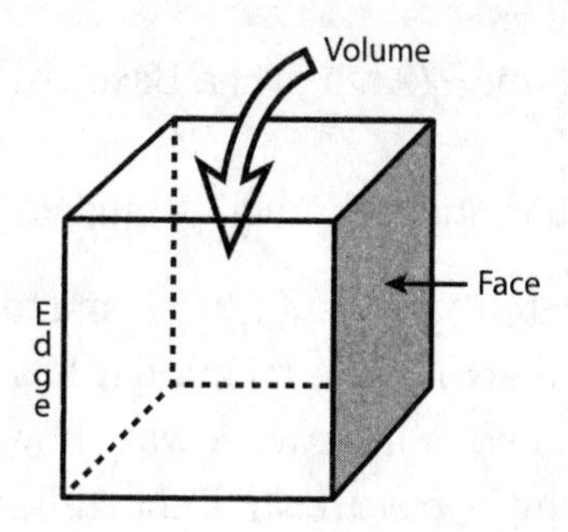

Since the total surface area is 150cm^2, then the area of each of the six congruent faces must be $\frac{150}{6}$, or 25 square centimeters. Since the area of each square is 25cm^2, then the length of each edge must be 5 centimeters. Now that you have the length of the edge, you can find the volume by cubing the 5. Therefore, $5 \times 5 \times 5$ equals a volume of 125 cm^3.

Volume is a third-dimensional measurement and is represented in units cubed, such as cm^3. *Area* is a second-dimensional measurement and is represented in units squared, such as cm^2. *Length* is a first-dimensional measurement and is represented in units such as cm.

Of course, the same question could be asked in reverse, giving the volume and asking for the total surface area.

If a similar question were to be asked using a more specific solid, such as a triangular prism, a lot more information would be given. It would be almost mandatory to sketch a picture of the object in question in order to make sense of it all. Although there is something to be said for the ability to do mathematics mentally, it increases the potential for mistakes, and what's worse, teachers won't be able to give partial credit if work isn't shown. A good rule is that you can never show too much work. For example:

> Find the prism height of a triangular prism with volume 600 cubic inches, whose triangular base is a right triangle with lengths 5 inches, 12 inches, and 13 inches.

See if you can find the mistake made in the following answer.

Prism Volume = (Area of the Base) × Prism Height

$600 = 78H$

The prism is about 7.7 inches tall.

Did you catch the mistake? The mistake made was using the wrong values to find the area of the triangular base. But that was extremely difficult to detect because so little work was shown!

By drawing a picture and showing a bit more work, not only will it increase the probability of getting the correct answer, it becomes a lot easier to follow:

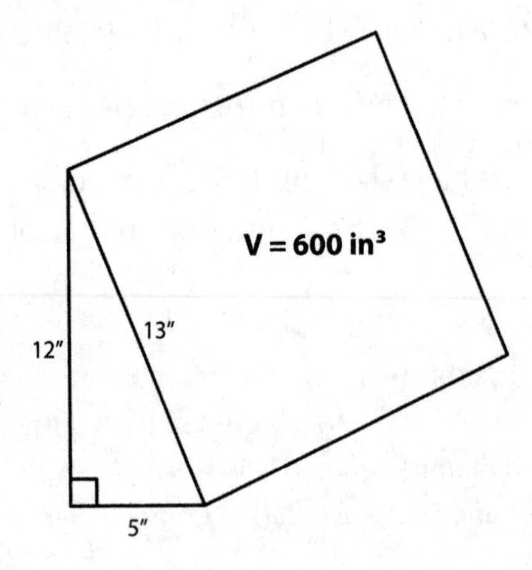

$$\text{Prism Volume} = (\text{Area of the Base}) \times \text{Prism Height}$$

$$600 = \frac{1}{2}(5)(12) \times H$$

$$600 = 30 \times H$$

$$600 \div 30 = (30 \times H) \div 30$$

$$20 = H$$

The prism is 20 inches tall.

Notice that the formula was written first before any specific numbers made an appearance. This is a good idea in any problem involving a formula. It truly is unbelievable how large of a role simple organization plays in the success of a student in mathematics. Organized work and decent penmanship will carry you a long way toward success in mathematics. If you think that your child's penmanship could stand some improvement, help him or her with it; their teacher will thank you.

Practice Problems

1. If 1 in = 50 ft, find the actual lengths being represented in the scale drawing.

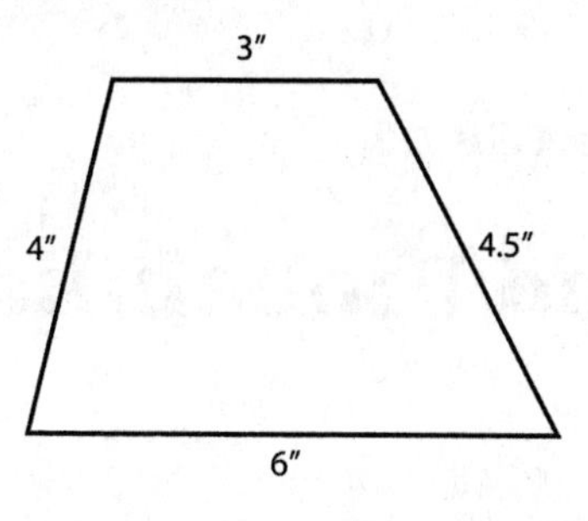

1 in : 50 ft

For the following two questions, use 3.14 for pi.

2. Find the circumference of a circle with a radius of 5 inches.
3. Find the area of a circle with a radius of 8 centimeters.
4. Find x:

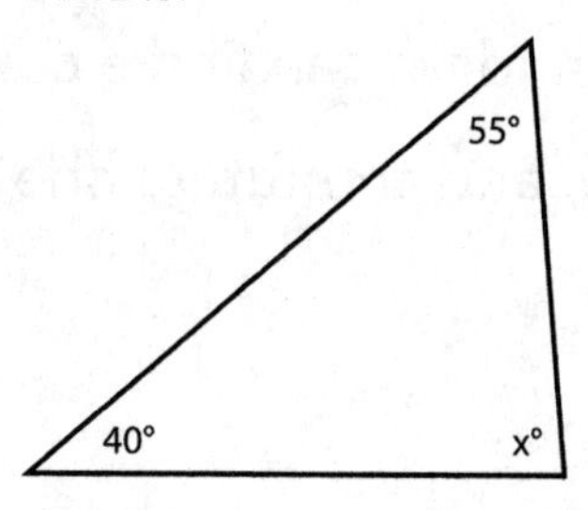

5. What angle measure is the supplement of 145°?
6. Find x:

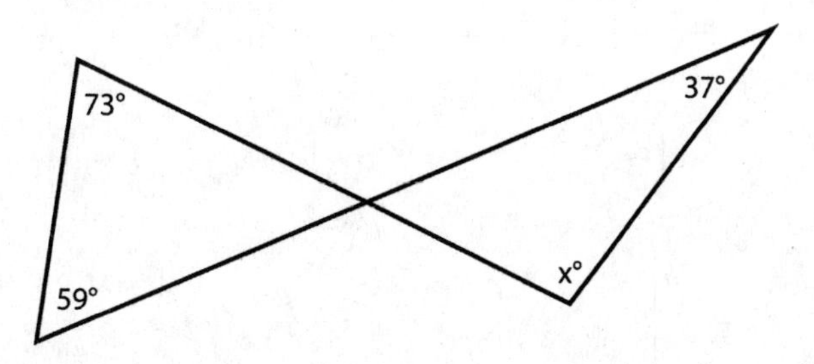

7. What is the total surface area of a cube whose volume is 64 cubic feet?

CHAPTER 15

Statistics and Probability

Statistics and probability help us make sense of the world. It is the bridge between social studies and mathematics—helping us learn from the past and improve upon the future. Whether it be meteorology, political science, or managing a big-league ball team, every student can find a connection between statistics, probability, and the future she is thinking about pursuing.

What Should My Child Already Be Able to Do?

Finding simple probability is a topic covered in the seventh grade. However, it is not uncommon for the material to be introduced as early as elementary school. Being fluent in the concept of probability allows for a much deeper discussion of statistical analysis in middle school. If you hear the word "probability," it may conjure images of rolling dice and spinning spinners. The truth is, the material will go much further than simple games of chance. Students *will* roll dice and spin spinners, but usually just to model other, more complicated, events.

The term *simple probability* refers to examples where one specific outcome is desired during one specific event. For example, trying to roll a five on a die. It does not necessarily mean a "very easy" probability problem. This differs from *compound probability*, where more than one event is happening. An example of this would be trying to roll back-to-back values on a die that are either odd or a four.

The basic *measures of central tendency* should be understood completely upon entering the seventh grade. These measures include mean (the average) and median (the central value within an ascending list). Oddly, the mode (the most frequent value) has been phased out of the CCSS. But it is not a complicated concept, so teachers may choose to discuss it during this section.

It is not a huge stretch to expect students to also be able to calculate range, first and third quartile values, and the interquartile range as incoming seventh graders.

Recall that the range is the difference between the maximum and minimum values, the quartiles are the values that occur 25% and 75% of the way through an ascending list, and the interquartile range is just the range of the quartiles. (The value at the 50% mark is the median.)

Even in the sixth grade, students use statistics to make inferences based on a set of information. They will have already begun using the measures of central tendency to describe a set of data. In the seventh grade, these statistics will be used to compare two different data sets and to make inferences regarding the differences and commonalities between them. For example, a sixth-grade student may be given a list of a football player's yards gained per game and then asked to explain what it means if the mean value is much greater than the median value. The seventh-grade student may be asked to compare the mean and median values from two different football players and make an argument as to who had the more successful season. Notice that although there is some interpretation within each of those examples, the seventh-grade question seems a bit more subjective. Mathematicians usually don't care for subjectivity in their work, so a new statistic, the *mean of absolute deviation* (MAD) will be utilized to make the jump from sixth- to seventh-grade statistical analysis (see Chapter 9 for more).

Random Sampling

Before we get into the mean of absolute deviation and how it is used, let's first go over some ground rules in statistics. Let's say somebody wanted to make some conclusions about a large collection of people (or any collection of things, for that matter). It would be impossible to gather information from each and every person or item in that population. A sample of that population would need to be used. However, the sample should be chosen carefully as not to skew any of the data.

As the old saying goes, "figures lie and liars figure." This is a shot at those who purposefully manipulate statistics either by how it is collected (improper sampling), or in how it is represented (misleading graphics).

As an extremely basic idea of this, imagine somebody is conducting a poll of New Yorkers' favorite food. The person plans on asking people what their favorite food is as they pass by on the street. If the person is standing in front of a popular pizza joint, with some of the people polled walking in and out of it, there is a good chance that the data is going to get skewed in the direction of pizza.

Polling needs to be done randomly, without bias. Considering different conditions that would lead to skewed data is one of the topics seventh-graders will cover in this unit. Some are obvious, as in the pizza example. Others are not so obvious. For example, a stadium employee asks the first thousand people who arrive at an evening baseball game what they think about the cost of food sold at that particular stadium. The data would not be a perfect representation of the entire fan base. Those eating dinner elsewhere before the game because they think the food costs too much are probably not going to be among the first thousand to show up. This may give the employee a false sense of how fans view the cost of the food at the stadium.

Mean of Absolute Deviation

As mentioned earlier, the mean of absolute deviation is a statistic used in numerical data sets. It cannot be used on categorical data such as eye color, nationality, or the previously mentioned favorite food. *Absolute deviation* can be found from either of two things: the median or the mean of the data set. Typically, it is found using the mean value of the set. Within all of our examples from here on, we will assume that the mean of absolute deviation is to be found using the *mean* of the data set, not the median.

There is a mean of the data set, and there is a mean of the absolute deviations. These are two completely different numbers. Do not get them confused with one another. Read carefully within the context of how it is being used to determine which mean is being mentioned.

The deviation of one particular value within the data set is how far away it is from the mean. Again, the absolute deviation is the *distance* between a number and the mean. This number cannot be negative. If your result of subtraction is negative, just ignore the negative sign. (For example, both of the values 4 and 22 would have an absolute deviation of 9 if the mean of the set was 13.) The easiest way to deal with this is to just start with the larger number when doing each of the subtractions.

Once the absolute deviation of every single piece of data in the set is found, the mean of those numbers is calculated. This value is the mean of absolute deviation. Let's look at a specific example from start to finish:

Find the mean absolute deviation of the following set of numbers.
4, 13, 7, 0, 12, 6, 18, 6, 9, 3

You first need to find the mean of the set. Recall that this is done by adding all of the numbers and dividing by how many numbers are present.

$$4+13+7+0+12+6+18+6+9+3=78$$
$$78 / 10 = \mathbf{7.8}$$

Now, you need to figure out the absolute deviation of every single number. (Remember, the difference cannot be negative.)

$$7.8-4=\mathbf{3.8}\ 13-7.8=\mathbf{5.2}\ 7.8-7=\mathbf{0.8}\ 7.8-0=\mathbf{7.8}\ 12-7.8=\mathbf{4.2}$$
$$7.8-6=\mathbf{1.8}\ 18-7.8=\mathbf{10.2}\ 7.8-6=\mathbf{1.8}\ 9-7.8=\mathbf{1.2}\ 7.8-3=\mathbf{4.8}$$

Now, it comes time to average all of those values.

$$(3.8+5.2+0.8+7.8+4.2+1.8+10.2+1.8+1.2+4.8) / 10 =$$
$$41.6 / 10 =$$
The MAD of this data set is **4.16**.

So what does the mean of absolute deviation tell us, anyway?
The mean of absolute deviation is very similar to the *standard deviation* that you may remember from statistics class in high school and/or college. It is a number used to measure how spread out the data set is. If the MAD is a relatively large number, it must mean that at least some of the numbers lie far away from the mean. If the MAD is a smaller number, the numbers must be bunched up, fairly close to the mean. Just how large that number needs to be in order to be considered "large" is a bit tricky. Therefore, MAD is usually used in middle school as a tool to compare two different (but similar) data sets.

Let's compare the MAD for two different running backs in football. The first athlete, Number 24, ran for 6, 4, 7, 12, 60, 5, 1, and 9 yards. The second athlete, Number 28, ran for 10, 15, 12, 20, 8, 11, 18, and 10 yards. After a little bit of quick addition, you can see that the running backs each ran eight times for 104 yards in the game. So who had the better day? It is not enough to go by total yards gained; they tied. And since they both ran eight times, their averages are exactly the same as well (13). Let's look a little bit closer at the data by calculating each of their MAD.

First, for Number 24:

$$13-6=\mathbf{7}\quad 13-4=\mathbf{9}\quad 13-7=\mathbf{6}\quad 13-12=\mathbf{1}$$
$$60-13=\mathbf{47}\quad 13-5=\mathbf{8}\quad 13-1=\mathbf{12}\quad 13-9=\mathbf{4}$$
$$(7+9+6+1+47+8+12+4)\,/\,8=$$
$$94\,/\,8=$$
$$\mathbf{11.75}$$

And now for Number 28:

$$13-10=\mathbf{3}\quad 15-13=\mathbf{2}\quad 13-12=\mathbf{1}\quad 20-13=\mathbf{7}$$
$$13-8=\mathbf{5}\quad 13-11=\mathbf{2}\quad 18-13=\mathbf{5}\quad 13-10=\mathbf{3}$$
$$(3+2+1+7+5+2+5+3)\,/\,8=$$
$$28\,/\,8=\mathbf{3.5}$$

It's not even close. Number 28's MAD is a lot lower than the MAD of Number 24. This may not clearly answer the question of "Who had the better day?" but it does give some ammunition for both sides of the debate. Maybe Number 28 had the better day because he was more consistent and reliable. Maybe it was Number 24 because he had the big 60-yarder. It is all in the interpretation.

Probability

It will become a lot easier to talk about probability once you review some important vocabulary. An *event* (at least in this sense) is something that occurs, which can result in a limited number of specific, measurable results. These results are called *outcomes*. A birthday may be an event in general, but it does not work as a mathematical event. Rolling a die is a perfect example, because it can only result in one of six known possible outcomes. The list of all possible outcomes is called the *sample space*. So, the sample space for a rolling a die is "1, 2, 3, 4, 5, 6." Conducting an event and recording the outcome is called a *trial*.

To eliminate any confusion between what qualifies as an event and what does not, some people refer to a mathematical event as a "chance event."

Okay, great. Now we can get into it.

Probability is the ratio between the number of possible desired outcomes compared to the number of total possible outcomes.

$$\text{Probability} = \frac{\textit{number of possible desired outcomes}}{\textit{number of total possible outcomes}}$$

The probability can be left as a fraction, or it can be converted and expressed as a decimal or a percent. The entire scale of probability happens between 0% and 100% as a percent, or 0 and 1 as a decimal. (To keep things consistent, probability here will be expressed as a decimal from now on.)

If something is *absolutely guaranteed* to happen, its probability is 1. Why? Well, let's say you roll a die and want a number less than ten to come up. The probability is $\frac{6}{6}$, or 1. By having every possible outcome also be a desired outcome, the ratio will always equal 1.

If there is *no chance* of something happening, then the probability is 0. Again, why? Well, if no desired outcome is possible, the ratio would be expressed as "0 over a number" and also equal 0. The probability of rolling an 82 on the die is $\frac{0}{6}$, or 0.

So, think about that for a moment: *All* probability is measured on a scale between two consecutive integers, 0 and 1. Negative probability does not make sense. Neither does a probability greater than one.

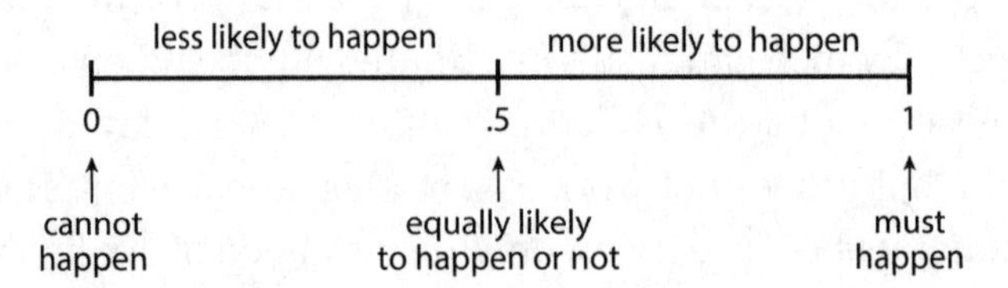

Why can't a probability be greater than 1?
In order for a probability to be greater than 1, the ratio would have to be an improper fraction. This would mean that there were more *desired* possible outcomes than *actual* possible outcomes. Simply put, that is impossible.

It may seem like a scale of 0–1 wouldn't be wide enough to cover all of the possible probabilities out there. What is easily overlooked is that you're working with decimals. And decimals can be extremely small but still greater than zero at the same time. Consider the probability of winning a fictitious multistate lottery. In order to win the jackpot, you have to choose five numbers from a pool of 1–60, and then another number from a separate pool of 35. What do you suppose the probability of winning is? One in a million? One in ten million?

The probability of winning the jackpot is one in 191,152,920! Written as a decimal, that would be approximately 0.00000000523. Plotting that point on our 0–1 number line would be practically impossible, seeing that its location would lie within the width of your pencil mark of the left-hand hash mark. Wow.

Experimental Probability

Probability, as we often think of it, is called *theoretical probability.* That is to say, "Theoretically, what are the chances of _____ happening?" There is another type called *experimental probability* (also known as relative frequency or empirical probability). This is where someone conducts trials a bunch of times, records the outcomes along the way, and makes inferences based on the results. For example, a student rolled a die 100 times and recorded the following frequencies: 10 1s, 14 2s, 16 3s, 24 4s, 11 5s, and 25 6s. The experimental probability of rolling a 6 would be $\frac{25}{100}$ (0.25). This value is much greater than the $\frac{1}{6}$ (or about 0.167) theoretical probability of rolling a 6 on a die. Students should realize that the discrepancy in the values is due to the relatively small sample size of 100 rolls. In time, you can expect the experimental probabilities to tend toward the expected, theoretical probabilities. This is called the *law of large numbers.*

Although the law of large numbers states that over extremely large amounts of trials, the experimental probabilities will closely represent the expected, theoretical probabilities, it is *not* accurate to utilize the concept as an immediate indicator of results. If tossing a coin amazingly resulted in twenty consecutive results of heads, there would *not* be a greater chance of the twenty-first toss turning up tails. It will have the same 50% probability as it always will.

Experimental probability has a very small role in mathematics. It *can* be used as a crude indicator of the probability of a specific outcome if mathematical means are not available, however. An example of this would be repeatedly dropping a thumbtack on the floor and recording how many times it lands point up. Generally, the probabilities discussed in middle school will all be able to be calculated to their accurate, theoretical value.

Modeling Events in the Classroom

Rather than asking a student what the probability of a particular outcome is, a teacher may ask a student to think of a chance event with a given probability. For example:

> The school plans on randomly choosing either the sixth, seventh, or eighth grade to get the newly renovated wing of the school. Using a die, construct a model to replicate the probability of your grade being chosen.

The die has a sample space of 6 outcomes. Since you are looking to model a $\frac{1}{3}$ probability, you must select two desired possible outcomes on the die. Any two will do; why not 3 and 6? Here's how a formally written answer would appear:

Event: Rolling a single die.

$$P(3 \text{ or } 6) = \frac{2}{6} = \frac{1}{3}.$$

Notice the notation of the probability. The P stands for "the probability of . . . " and then whatever is in the parentheses is the desired outcome.

It is crucial that the specific event is given in a problem, as well as the desired outcome. It is not enough information to ask what the probability of "getting a 5" is if there is no context to assess the sample space of the event.

After creating the model, students may be asked to run trials repeatedly to verify whether or not the experimental probability is correctly representing the desired theoretical probability. This seems a bit exhausting, but there are many software programs that automatically conduct events such as rolling a die, tossing a coin, and spinning a spinner. They can be set to conduct as many trials as you tell them to do, sparing the students from sore fingers and the entire class from ringing ears.

Compound Probability

Compound probability is calculated when more than one event is happening. Sometimes the multiple events occur simultaneously, as in rolling two dice. Sometimes they happen in succession, as in flipping over two cards from a standard deck.

Calculating compound probability is done by finding the individual probabilities and then multiplying them together. For example:

A 1–10 spinner is spun twice. What is the probability of spinning a number greater than seven twice?

The probability of the first winning spin is $\frac{3}{10}$. And there is no reason for that probability to be any different for the second spin. Therefore, the entire probability for back-to-back winning spins is:

$$\frac{3}{10}\left(\frac{3}{10}\right)=\frac{9}{100} \text{ or } 9\% \text{ or } 0.09$$

When working with compound probability, it is important to consider the possibility of the second event being affected by the first event's outcome. Such cases are called *dependent events*. An example would be drawing two cards without replacing the first card.

Let's look at another example, this time involving dependent events:

What is the probability of choosing a face card and then an ace from a standard deck of playing cards, if the first card is not replaced?

There are 52 cards in a deck, 12 of them being face cards (jack, queen, and king in each of the four suits). The probability of the first desired outcome is $\frac{12}{52}$. This fraction can be simplified to $\frac{3}{13}$, but it might be a good idea to hold off on simplifying the fraction for a moment. The original fraction will sometimes help us find the second probability.

Part two of the compound event asks for an ace. Remember, you already chose one card . . . but you assume success to continue. You chose a face card, so all four aces are still in the deck. However, there are only 51 cards left. Here's how that looks without the long description:

Event: Choosing two cards without replacing.

$$P(\text{Face card then ace}) = \left(\frac{12}{52}\right)\left(\frac{4}{51}\right) = \frac{48}{2652} = \frac{4}{221} \approx 0.018$$

Notice that by leaving the 52 in the first denominator, you were able to subtract the one chosen card and arrive at the second denominator of 51. This delaying of the simplifying is especially helpful in a compound probability problem, such as the lottery problem from earlier where five numbers had to be chosen from 1–60 and then another from 1–35:

$$P(\text{Winning ticket}) = \left(\frac{5}{60}\right)\left(\frac{4}{59}\right)\left(\frac{3}{58}\right)\left(\frac{2}{57}\right)\left(\frac{1}{56}\right)\left(\frac{1}{35}\right) =$$

$$\text{Simplify the 5 \& 60 and the 3 \& 57: } \left(\frac{1}{12}\right)\left(\frac{4}{59}\right)\left(\frac{1}{58}\right)\left(\frac{2}{19}\right)\left(\frac{1}{56}\right)\left(\frac{1}{35}\right) =$$

$$\text{Simplify the 4 \& 12 and the 2 \& 58: } \left(\frac{1}{3}\right)\left(\frac{1}{59}\right)\left(\frac{1}{29}\right)\left(\frac{1}{19}\right)\left(\frac{1}{56}\right)\left(\frac{1}{35}\right) = \frac{1}{191{,}152{,}920}$$

or about 0.0000000523.

The order of the chosen numbers does not matter, so you start with hoping for any of the five numbers on the ticket out of a possible sixty. After that, assuming success along the way, you have one less desired number left on the ticket and one less number left in the machine . . . until the last number, which there is one number out of 35 that you need.

"1 out of 191,152,920" is tough to wrap the mind around. Picture it this way: That's the same ratio as one second in time as compared to 6 years, 21 days, 10 hours and 2 minutes!

The math involved in that example was a lot to keep track of, so leaving the fractions unsimplified in the first line of work shown was a good idea. Obviously, after that, go ahead and attack that long string of multiplication, simplifying whatever you can to help whittle down the size of the numbers.

Tree Diagrams and Organized Lists

Certainly not in the case of the previous lottery problem, but sometimes, it is possible (and visually easier) to draw out the entire sample space of compound events. The most straightforward way to do this is with a tree diagram.

Let's say that a fair coin (that is to say, a coin which is equally as likely to come up "heads" as "tails" . . . no funny business!) is going to be tossed 3 times. What is the sample space of the entire compound problem? Picture each toss of the coin as a fork in the road. The coin can either come up heads or tails, and you can either go left or right. Starting at a point on the left and drawing a fork at every possible step along the way, you get a structure like this:

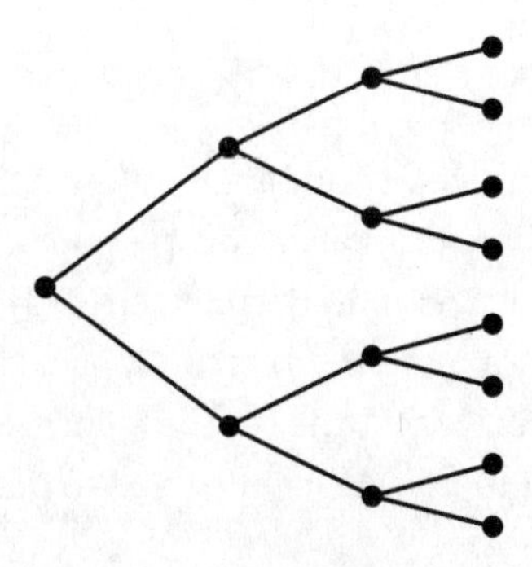

But now, instead of "left or right," label the diagram with "heads and tails" at each fork:

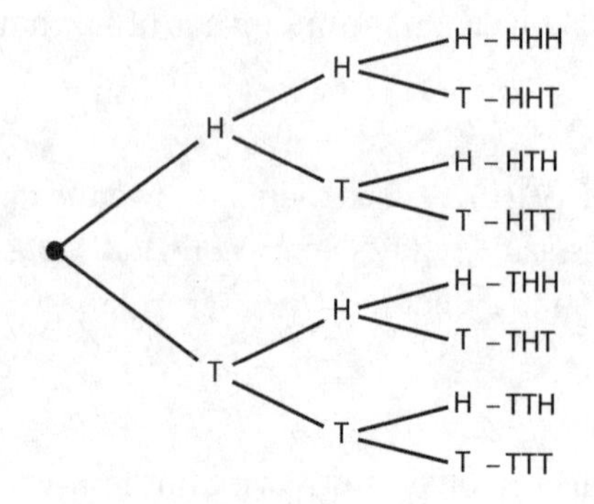

The sample space is represented by all of the final possible "destinations" on the right-hand side of the diagram.

It makes sense that there are eight possible outcomes for the entire event. That's because three distinct, independent events occurred—each with two possible outcomes. $2 \times 2 \times 2 = 8$.

How many ways can athletes A, B, C, and D finish a race?

You *could* draw another tree diagram, but that may start to get unruly, as there are now four things in play. You could also make an organized list—which you will—after you reason with the problem a bit first.

From the beginning, there are four different people who could win the race. It does not make any difference who crosses the finish line first; there will then be three people competing for second. Then two people will complete for third place. And finally, the last person will cross the line (we hope). The number of possible final outcomes can be calculated by:

$$4 \times 3 \times 2 \times 1 = 24 \text{ possible outcomes}$$

Calculating the total number of ways that items can be arranged falls in a field of mathematics called *combinatorics*. There are two major categories of ordering items: permutations (where order matters) and combinations (where order does not matter). There are specific formulas for calculating each of these, but the work in the seventh grade will be done with numbers small enough that there will be little need to get into memorization of formulas.

If asked to give the sample space of that event, the entire list of 24 possibilities would need to be written out. The best way to attack that task is with an organized list. Work slowly and develop a pattern that works best for you. Usually, the best way to organize a list like this one is alphabetically.

As promised, here is the organized list of the 24 possibilities:

ABCD	BACD	CABD	DABC
ABDC	BADC	CADB	DACB
ACBD	BCAD	CBAD	DBAC
ACDB	BCDA	CBDA	DBCA
ADBC	BDAC	CDAB	DCAB
ADCB	BDCA	CDBA	DCBA

That may have felt like a bunch of busy work, but having the sample space in front of you can make some questions a whole lot easier to answer. For example:

> How many ways could the race end where runner B finishes ahead of runner C and runner A finishes ahead of runner D?

Since you have the list available, just highlight the cases where that holds true:

ABCD	**BACD**	CABD	DABC
ABDC	**BADC**	CADB	DACB
ACBD	**BCAD**	CBAD	DBAC
ACDB	BCDA	CBDA	DBCA
ADBC	BDAC	CDAB	DCAB
ADCB	BDCA	CDBA	DCBA

There are six ways that those conditions could result at the conclusion of the race.

If you didn't have the list as a resource, that question would have been brutal to answer. Whenever possible, make the problem easier for yourself by either drawing a picture or making an organized list.

Practice Problems

1. Find the mean of absolute deviation for the following set of data:

   14, 16, 16, 18, 19, 20, 22, 24, 27, 29

2. What is the sample space for the event of spinning a 1–10 spinner?
3. If a desired outcome has a probability of 0.99, is it very likely or very unlikely to happen?
4. If a penny is tossed 29 times and heads comes up every time, what will the probability of tails coming up on the 30th toss be?
5. Two cards are chosen from a standard deck. What is the probability of an ace being chosen, followed by *either* a ten or a face card?
6. Mr. and Mrs. Jones plan on having two children. Draw a tree diagram to show the possibilities of the two children's genders.
7. How many different ways could a complete 8-horse race finish? Do not list them all.

PART 4

8th Grade Common Core Standards

CHAPTER 16

What Is Expected of My 8th Grader?

To answer the question posed by the title of this chapter in two words—a lot. Although eighth graders are not yet in high school, it will, at times, certainly feel as if they are. Some of the math is very abstract. Long gone are the lessons where numbers represented amounts of tangible things. This is where the math starts to get really abstract.

What Are the Critical Areas?

There are three areas in which eighth-grade students will be spending the majority of the school year. These areas are defined by Corestandards.org as:

1. Formulating and reasoning about expressions and equations, including modeling an association in bivariate data with a linear equation, and solving linear equations and systems of linear equations
2. Grasping the concept of a function using functions to describe quantitative relationships
3. Analyzing two- and three- dimensional space and figures using distance, angle, similarity, and congruence, and understanding and applying the Pythagorean Theorem

Wow. Just reading *what* will be covered was exhausting. Luckily, students and teachers have an entire school year to get through it all. Remember, that was part of the idea behind the core standards in the first place: depth, not breadth. Those concepts may have sounded intense, but there are really only three of them. Even if you break them down into the individual topics that the students will see throughout the year, those topics are so interrelated that it will not feel as overwhelming as you may think right now.

Critical Area One

Again, the main focus of the math in eighth grade gravitates around linear relationships. To break that broad topic down further, you need to think about a relationship mathematically. Luckily, we are not discussing the kind of relationships in eighth grade that start with a cute note from Sally to Billy and ultimately end up with Sally crying and Billy confused about what he did wrong. We are talking about relationships between variables, inputs and outputs, cause and effect.

The collection of input values is called the *domain*. Operations then happen to that input. There may be only one operation, or there may be many. The results of these operations on the inputs then become the output values. Collectively, they are called the *range*.

There is more than one place where the word "range" is used in math—and they have extremely different meanings. Another range, the one used in statistics, means the difference between the maximum and minimum values in a data set. Although somewhat similar, that meaning is not directly associated with this concept.

The word "linear" means relating to or resembling a line, and refers to the graph produced. *Linear* relationships produce straight lines when inputs and their corresponding outputs are plotted on the Cartesian plane.

The Cartesian plane is just a fancy name for the coordinate plane. It is named after Renè Descartes who was a mathematician and philosopher in the early seventeenth century. He was a pioneer in the bridgework between geometry and algebra.

This may happen as a specific set of given individual points that fall on a straight line, or as an *infinite* set of points that form a complete straight line. In order for a relationship to be linear, the only operations that can happen to the input are addition, subtraction, multiplication, and division, and they have to be with known numbers only—not other variables.

From the word "relationship," we get the word relation. A *relation* is simply a set of ordered pairs. It is possible for the relation to be expressed as an equation, as long as there are two separate variables present. One variable will be defined as the input (usually x) and the other the output (usually y), and ordered pairs of the relation can be generated from the equation.

With equations that only have one variable, you can't possibly get an ordered pair as a solution. Ordered pairs are expressed as (x, y) and are plotted on the Cartesian plane. Those equations with only one variable will have solutions that can be plotted on a number line instead.

In the world of statistics, ordered pairs are known as *bivariate* data. All of the same rules about linear versus nonlinear still apply, even though there may not be a perfect equation tying together the input and the output.

The biggest difference here will be that students are not given a formula ahead of time to find the output values. They will run experiments and surveys to get a collection of data in ordered pair form. They will then plot their ordered pairs and look to see if the points fall along a straight line or not. If so, they will be able to generate a linear graph that serves as a representation of the data they collected. This math is not going to be a perfect science, so it will be important for students to be able to use their best reasoning skills to draw conclusions and make connections within the mathematics they are studying.

Systems of Equations

A lot of work will be done exploring the different ways that bivariate equations can be represented: tables, graphs, and words, as well as the equation itself. However, it will be much more beneficial to students to be able to compare one equation with another. When two (or more) equations are viewed within a problem at the same time, you have a system of equations.

In a system of equations, you will see variables more than once. Although you begin the problem not knowing their values, you do know that whatever the value of one variable is, every one of that same variable will be worth the same thing.

The basic idea behind solving a system of equations is to find the values of the variables that solve not just one equation, but every equation in the system. Students can arrive at the solution three different ways. The Common Core Standards are a bit vague about how far teachers should go with this topic. Up until they came out, solving systems of equations was classified as Algebra I material, and all three methods were covered extensively. The Standards merely state that "students will solve a system of equations algebraically." That would be true of all three methods. The level of difficulty for these problems can range from introductory to an intense complexity.

$y = x + 6$ $\quad\quad$ $3x - 5y = -38$

$y = 10$ $\quad\quad$ $2x + 4y = 48$

Both of these systems have a solution of $(x, y) = (4, 10)$ but vary greatly in difficulty.

Teachers will differentiate their instruction to push students to their maximum potential. Some students will not be exposed to the same problems as others. Some classes may not even be expected to know all three ways to solve a system of equations. Knowing how far to go with the material is up to the teacher. They have been well educated in this. Do not feel as though your child is being shortchanged if you do not see a lot of this material come home with him.

Critical Area Two

For those students who continue with math for all four years of high school, as well as at college or university, the concept of functions is going to stick around for a long, long time. Obviously it's a very important topic, but interestingly enough, by the time it is introduced, students will have already done the majority of the work necessary to understand what is going on.

Students should, at this stage of the game, understand the concept of performing operations to an input value to get an output value within an equation. The shift from working with equations to working with function notation is a seemingly obscure one. Instead of using the familiar y as the output variable, $f(x)$ is used. This new symbol stands for "the function applied to the input variable." What this does is allows you to look more closely at each individual input/output relationship, using the equation as more of a formula. For example, the following function can be used to convert temperature measured in degrees centigrade into Fahrenheit:

$$f(x) = \frac{5}{9}x + 32$$

If we wanted to know what 36°C is in on the Fahrenheit scale, we wouldn't necessarily be concerned with what the graph looked like. We'd only care about $f(36)$.

$$f(36) =$$

$$f(36) = \frac{180}{9} + 32$$

$$f(36) = 20 + 32$$

$$f(36) = 52$$

$$36°C = 52°F$$

The previous example may happen to be a linear function, but students may find themselves getting away from the restriction of linear relationships once function notation shows up. However, that is of little concern. The overall pattern that a nonlinear graph makes doesn't play a large role at this stage in function notation. Save that for high school. All this means is that students will be asked to perform some more elaborate operations in order to get individual output values. For example, the graph of $y = x^2 - 4x + 3$ may look very intimidating:

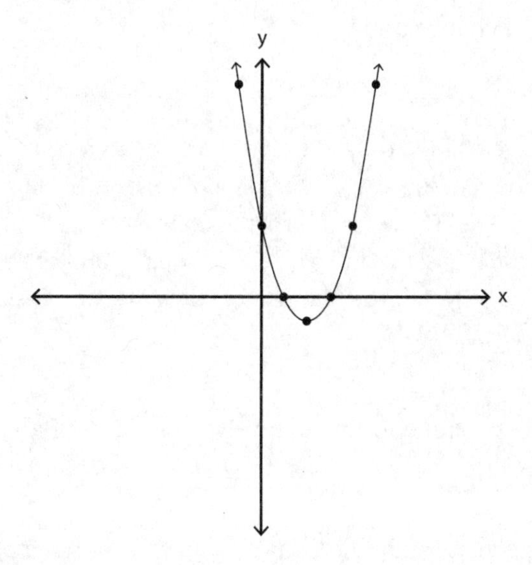

But finding $f(2)$ is very doable:

$$f(2) = 2^2 - 4(2) + 3$$
$$f(2) = 4 - 8 + 3$$
$$f(2) = -1$$

The shape of the previous function is called a *parabola*. It appears when the input variable is being squared. Instead of a straight line, it is constructed with a smooth curve.

The shape of these more elaborate functions will serve as the focus in Algebra II and Calculus.

Critical Area Three

Although critical areas one and two are very much related to one another, critical area three is quite different. Students will spend this time exploring two-dimensional shapes and three-dimensional solids. The only connection this last critical area has is that it involves a lot of formulas in order to find areas of shapes and volumes of solids.

We've pointed this out before, but it bears repeating: Very often, people will incorrectly identify a three-dimensional object such as a cube or a sphere as a "shape." Those are not shapes; they are actually called solids. Shapes are two-dimensional figures, such as squares and circles. This error is made so frequently that you may even catch a math teacher making it!

Most volume formulas involve some use of an area formula. Students will be expected to have at least the basic area formulas memorized. These include:

- Area of a rectangle/square

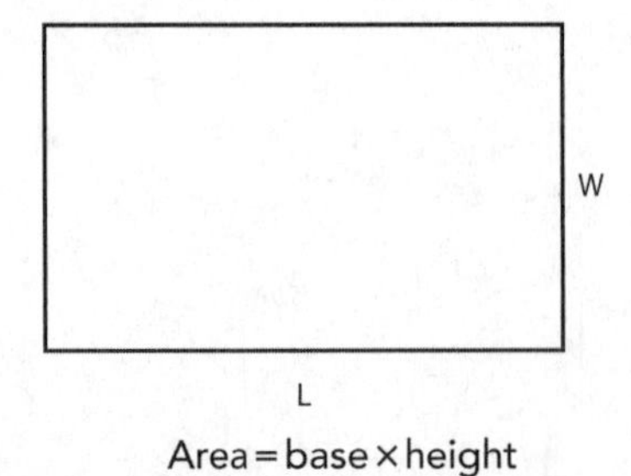

Area = base × height

- Area of a triangle

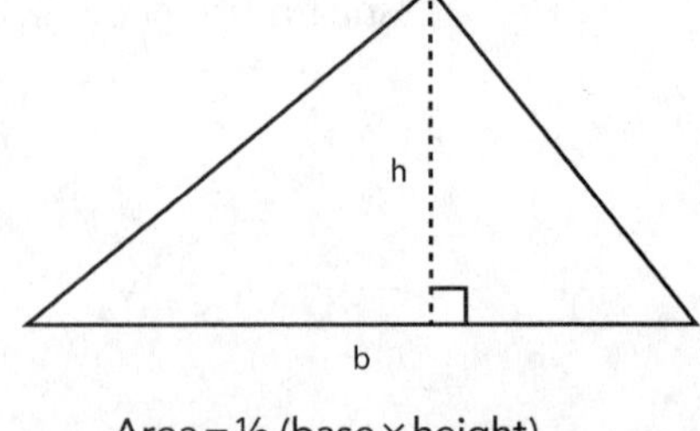

Area = ½ (base × height)

It may sound silly but students can actually prepare for this unit ahead of time by practicing drawing the following solids.

- Rectangular Prisms

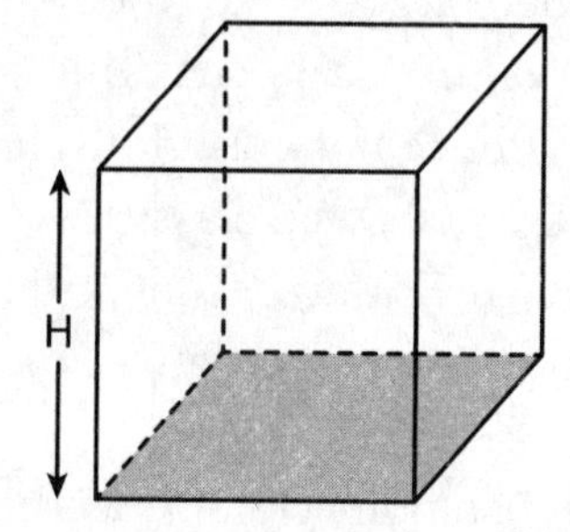

- Triangular Prisms

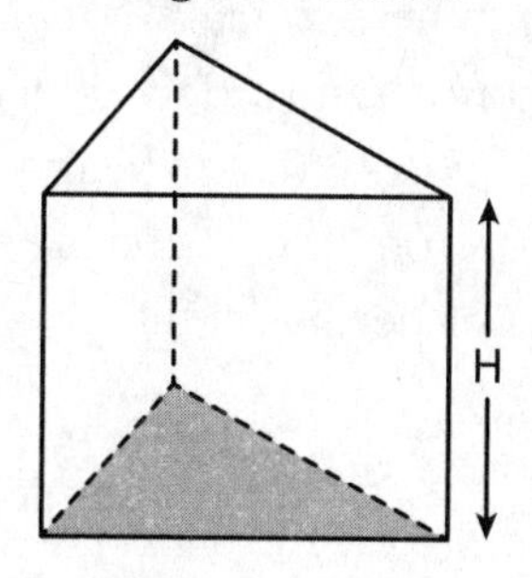

- Cylinders

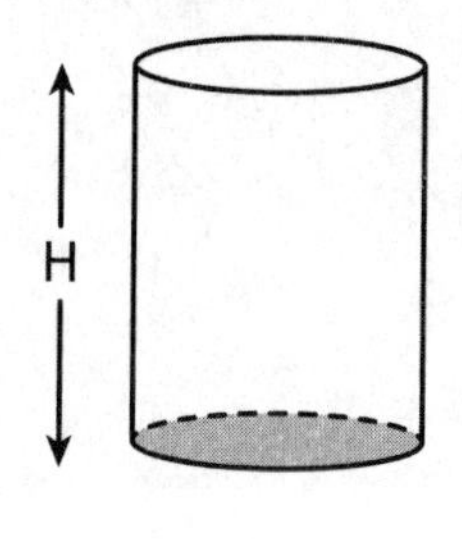

- Cones

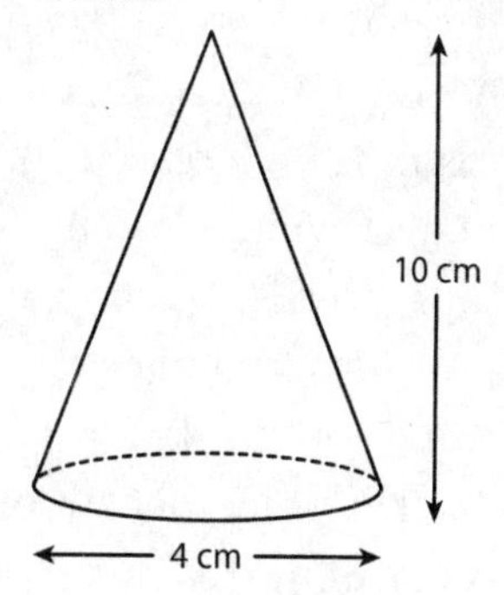

- Pyramids

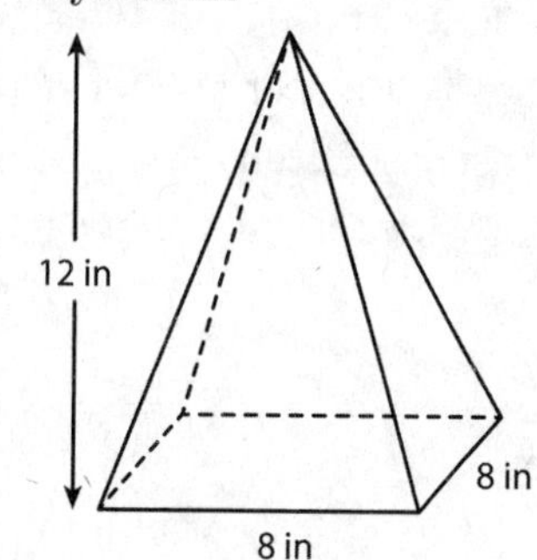

- Spheres

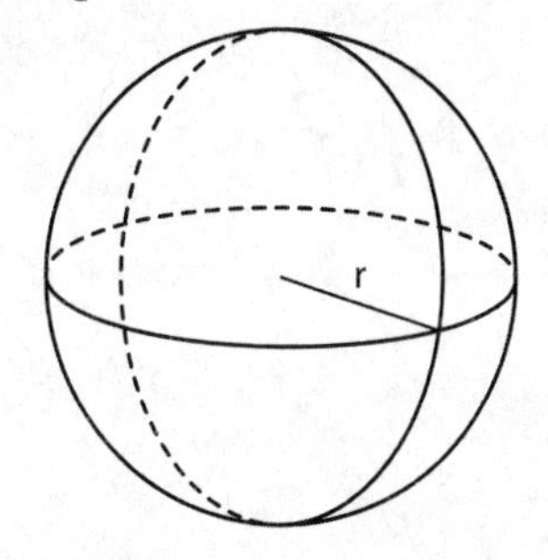

Often, students who struggle with this unit do so because they have a difficult time interpreting the diagrams of the three-dimensional solids. By practicing drawing the designs ahead of time, students will grow more familiar with the individual components of each solid. Once they can identify where to find the essential parts of the solids, answering the questions becomes as easy as plugging numbers into a formula.

The Great Math Debate

There is some controversy among teachers as to whether or not students should have the surface area and volume formulas for two- and three-dimensional shapes and solids memorized or not. Be ready for either decision with your child's teacher. Again, this is something that is not explicitly addressed in the Common Core Standards. Your child's teacher may provide a formula sheet for all students, some classes, some students, no students . . . it's impossible to predict. However, standardized tests do provide a formula sheet to students, so although you cannot be certain, you can be optimistic.

The same debate exists about the use of calculators within the classroom. Some teachers argue that the use of calculators diminishes the automaticity of students' basic facts. Other teachers claim that by forgoing longhand calculations, students have more time to delve deeper into the newer material. Sometimes, schools as a whole have policies regarding both issues. Often, it is left up to the teacher. Bottom line is, choose either side of the debate as you wish—just be ready for either possibility.

CHAPTER 17

The Number System

When you first started to learn about numbers, you were exposed to 1, 2, and 3. In time, you learned more and more numbers, and people asked you how high you could count. At some point, zero was introduced as the number to express having none of an item. The next major revelation was that there are partial amounts of numbers called either fractions or decimals. You then welcomed values less than zero, the negative numbers. That concluded the real number system. The final set of numbers, complex numbers, will not come into play until Algebra II when square roots of negatives are explored. This chapter will concentrate on the real number system.

What Should My Child Already Be Able to Do?

Although it does not guarantee success in the eighth grade, one of the most important factors in students being able to learn new material is their automaticity of basic arithmetic facts. If your student cannot recite multiplication facts for integers 0–9, make flash cards. Start practicing now . . . and often. If he cannot tell you that the sum of 7 and 6 is 13 without pausing to think about it, quiz him during dinner or a game of catch.

It is very difficult to be able to concentrate on absorbing new material if, for example, the first step of many is to multiply 7 by 8, and a student is slowed down right out of the gate.

Sign rules (the arithmetic of positive and negative numbers) can be tricky because they are so easily confused with one another. The easiest way to work with sign rules is to break them into two major categories:

1. Multiplying and dividing
2. Adding and subtracting

The result of multiplying or dividing any two positives *or* any two negatives will *always* be positive. The result of multiplying or dividing one positive and one negative (regardless of size or order) will *always* be negative.

$$4 \bullet -7 = -28 \text{ and } -4 \bullet 7 = -28\text{, but } -4 \bullet -7 = 28$$

The rules for adding and subtracting are a bit more complex because the magnitude of the number (its distance from zero) comes into play when figuring out if the result is going to be positive or negative. Furthermore, the order of the numbers when subtracting is also significant. Fortunately, you can always transform any subtraction problem into an addition problem by switching the sign of the second number.

$$15 - (-28) =$$
$$15 + (+28) = 43$$

The sum of any two positives will *always* be positive—add the magnitudes to figure the answer. The sum of any two negative will *always* be negative—add the magnitudes to figure the answer. The sum of opposite signs will have the same sign as the number with the larger magnitude. *Subtract* the magnitudes to find the number attached to that sign.

$$-9+4=-5$$

The magnitude of the first addend is nine, the second is four. Since there are five more negatives than positives, the answer is negative five.

Most of the number system will be covered by the time that students reach the eighth grade. The less that they have to review, the more enjoyable and successful their experience will be throughout the more complicated material during the last year of middle school.

Rational versus Irrational

The root word of rational is "ratio." A *ratio* is a comparison of two numbers. A *fraction* is a comparison of two numbers. So, summed up quickly, a rational number is any value that can be expressed as a fraction.

But not always. By that definition, $\frac{\sqrt{2}}{\pi}$ would be rational . . . but it is not. Before you define exactly what a rational number is, let's recall that an integer is any number, positive or negative, that has no fractional or decimal part.

$$-6, 3, 0, 729 \text{ are all integers.}$$

$$8.2, \frac{3}{4}, \pi \text{ are all } \textit{not} \text{ integers.}$$

A rational number is any value that can be expressed as a fraction, *a*/*b* where *a* and *b* are both integers and $b \neq 0$.

Some numbers are very easy to identify as rational. $\frac{1}{2}$ is obviously rational because it clearly fulfills the criteria. 0.25 is rational because it is the same as $\frac{1}{4}$, and that also abides by the rule.

Some numbers are not as obvious. 0.82? π? −27? We need a bit more clarity.

Figures such as square roots and fractions should be evaluated and simplified before they are classified into a subset of numbers. Otherwise, a number like $\frac{32}{4}$ may be overlooked as being an integer!

Simplified and evaluated numbers will all fall under *exactly one* of three subsets: integers, decimals, and fractions. We have already discussed that your typical, no-funny-business fraction is rational. Any integer *must* be rational because it could simply be written as the fraction of itself over 1.

This leaves the decimals.

A decimal is considered rational if it either terminates (stops at some point) or repeats.

▼ EXAMPLES OF RATIONAL NUMBERS:

$\frac{1}{14}$	−2	0	0.14	$.\overline{7}$	$-\frac{5}{19}$

This leaves only non-repeating, non-terminating decimals. These numbers are called *irrational*. Some irrational numbers can be created from strange patterns; 0.01001000100001 . . . is irrational. (There is no perfectly repeating pattern, but it doesn't ever terminate, either.)

There are much more common examples of irrational numbers. π (pi) is probably the most famous irrational number. We use 3.14 as an approximation, but the exact decimal could never be written out; it never ends!

People sometimes compete to see who can memorize the most digits of pi. On November 20, 2005, Chao Lu of China set the world record in digits memorized at an astonishing 67,890!

Additionally, any square root that does not "work out nicely" is an irrational number.

Square Roots

A square root (also called a *radical*) of a number is the value that would have to multiply by itself in order to get back to the original number.

$\sqrt{16} = 4$ because $4 \bullet 4 = 16$.

$\sqrt{100} = 10$ because $10 \bullet 10 = 100$.

Sometimes square roots work out "nicely" and sometimes they do not. "Working out nicely" just means that the result is, at worst, a fraction. "Working out rationally" is a much more technically sound phrase, but "working out nicely" is often used instead.

Since it worked out rationally, $\sqrt{16}$ is a rational number.

$\sqrt{2} = 1.4142135623$ . . . blah blah blah. It does not work out "nicely," and therefore is irrational.

It is very important to note that square roots of negatives are not allowed until Algebra II, when the last subset of the entire complex number system, the "imaginary numbers," is introduced into the mix.

Why does a calculator give an error message when $\sqrt{-4}$ is typed?
A square root is the number that must be multiplied by its *exact* self. No number when multiplied by itself results in a negative value. Therefore, taking the square root of a negative is "undefined," and angers calculators.

Approximating Square Roots

Since irrational square roots do not work out to exact numbers that you can write out with complete accuracy, you have three choices—leave it written as a square root (this is called radical notation), estimate it, or round it.

To estimate a square root, it is helpful to have the first few perfect square numbers memorized.

▼ PERFECT SQUARES 1–400

$1^2=1$	$2^2=4$	$3^2=9$	$4^2=16$	$5^2=25$	$6^2=36$
$7^2=49$	$8^2=64$	$9^2=81$	$10^2=100$	$11^2=121$	$12^2=144$
$13^2=169$	$14^2=196$	$15^2=225$	$16^2=256$	$17^2=289$	$18^2=324$
$19^2=361$	$20^2=400$				

You can now clearly see that $\sqrt{78}$ is between 8 and 9. We can go a bit further to say that it is a little less than 9, since 78 is closer to 81 than it is to 64.

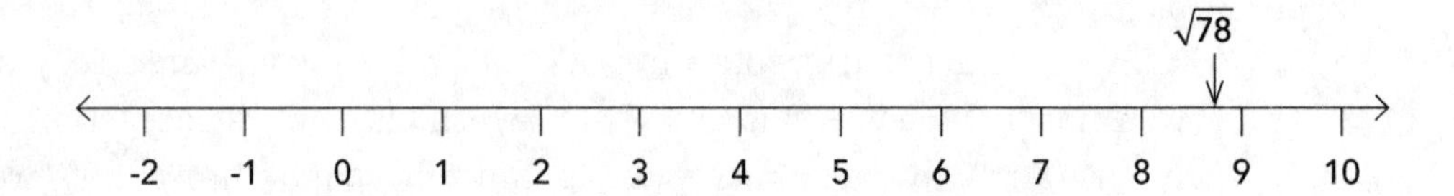

You can estimate $-\sqrt{21}$ to be between -4 and -5.

$-\sqrt{21}$ means "the opposite of the square root of 21" and is allowed. $\sqrt{-21}$ means "the square root of -21" and is not allowed at this stage. Remember, square roots of negatives are not recognized until Algebra II.

If a bit more precision is desired, square roots can also be rounded. In order to evaluate a square root, you are going to need a calculator. Punch $\sqrt{21}$ into a calculator and you will get 4.5825757. (The actual number continues on forever in a non-repeating fashion, but the calculator must round it off.)

- If asked to round $\sqrt{21}$ to the nearest thousandth, you'd get ≈ 4.583.
- If asked to round it to the nearest hundredth, you'd get ≈ 4.58.
- If asked to round it to the nearest tenth, you'd get ≈ 4.6.

Practice Problems

1. Which of the following are rational numbers?

   $-4,\ \pi, \sqrt{2}\ 3.6,\ 0,\ \sqrt{81},\ \frac{-12}{7}$

Find the value of each.

2. $\sqrt{36}$
3. $\sqrt{33}$
4. $-\sqrt{9}$

Between which two consecutive integers are each of the following?

5. $\sqrt{64}$
6. $\sqrt{130}$
7. $\sqrt{170}$

Use a calculator and evaluate each of the following to the nearest hundredth.

8. $\sqrt{76}$
9. $\sqrt{215}$
10. $\sqrt{500}$

CHAPTER 18

Expressions and Equations

Even before the Common Core showed up, there was a huge emphasis on algebra in the eighth grade. This is the year when the alphabet starts rearing its head in the form of variables. They show up early and often. This is a fairly long chapter because, quite frankly, there is an incredible amount of material to cover. Variables can be considered helpful because they are so versatile. However, they can also be the bane of students' mathematical careers because they signal the beginning of abstract concepts—when math is no longer about the numbers you can see right in front of you. Sit back, relax, and take a deep breath; it's not as bad as people make it out to be.

What Should My Child Already Be Able to Do?

One of the biggest changes to the middle-school math curriculum is the timing of when students are introduced to solving equations. Students should go into the eighth grade knowing how to combine like terms and how to solve basic equations.

There will be some work involving exponents, but it will be expected for students to already have a basic understanding of what it means to raise a number to an exponent. The eighth-grade exponent work will revolve around patterns, properties, and practicalities.

Students will be asked to "undo" some exponents within equations. This won't get crazy, but it will be important for them to know what it means to take the square root of a number.

Taking the square root of a number feels a *lot* like division . . . even the symbol $\left(\sqrt{\ }\right)$ is reminiscent of the old division setup. A common mistake when trying to find a square root is to halve a number instead.

With as much work with proportions as there was in the seventh grade, it seems nearly impossible for there to be anything new to cover. Oh, but there is. All of those proportions covered in the seventh grade are now going to be represented graphically. Contrasts, comparisons, and connections are going to be made, so a solid understanding of proportions is imperative.

Integer Exponents

Yes, it is important to specify the exponents covered in the eighth grade as being integers, because there actually is such a thing as a fractional exponent. The good news is that fractional exponents are not in the eighth-grade curriculum.

Zero and Negative Exponents

It sounds weird but there is such a thing as raising a number to a negative power. You can even raise a number to the power of zero. The easiest way to see how it works is to make your way down the following table:

▼ POWERS OF 4

$4^3=4\bullet4\bullet4=64$	
$4^2-4\bullet4-16$	(previous line divided by 4)
$4^1=4$	(previous line divided by 4)
$4^0=1$	(previous line divided by 4)
$4^{-1}=\frac{1}{4}$	(previous line divided by 4)
$4^{-2}=\frac{1}{16}$	(previous line divided by 4)
$4^{-3}=\frac{1}{64}$	(previous line divided by 4)

There's a really cool connection here. See it? $4^3=64$ and $4^{-3}=\frac{1}{64}$. Whenever you see a negative exponent, it can be made positive by taking the reciprocal. If the negative exponent is *already* in the denominator, taking the reciprocal will send it to the numerator.

$$2^{-4}=\frac{1}{2^4}=\frac{1}{16}$$

$$\frac{3}{2(5^{-2})}=\frac{3(5^2)}{2}=\frac{3(25)}{2}=\frac{75}{2}\text{ or }37.5$$

In the previous example, the 2 in the denominator did not move because it was not part of the exponent's base. The exponent of −2 only affects the 5.

Additionally, you can see that raising a value to the zero power results in 1. This is because an exponent of 0 means "none of that base." Since exponents deal with multiplication, only the multiplicative identity, 1, remains.

Properties of Exponents

Exponents signify repeated multiplication. So, if two exponential expressions with the same base are multiplied together, then the exponents get added together.

$$a^3 \bullet a^4 =$$
$$(a \bullet a \bullet a) \bullet (a \bullet a \bullet a \bullet a) =$$
$$a \bullet a \bullet a \bullet a \bullet a \bullet a \bullet a =$$
$$a^7$$

The bases must be the same to be able to combine exponential expressions together. The expression a^3b^4 cannot be simplified any further.

Now if you look at the same thing in reverse, you can make another solid connection. If multiplying common bases means *adding* the exponents, then dividing common bases must mean:

$$\frac{a^8}{a^2} =$$

$$\frac{a \bullet a \bullet a \bullet a \bullet a \bullet a \bullet a \bullet a}{a \bullet a} =$$

$$a \bullet a \bullet a \bullet a \bullet a \bullet a =$$

$$a^6$$

Yep, you guessed it—you subtract the exponents. Or, put more elegantly:

$$\frac{a^8}{a^2} = a^{(8-2)} = a^6$$

This last property is pretty neat—if an exponential expression is raised to *another* power, then the repeated multiplication gets repeated itself and you multiply the exponents.

$$(a^4)^3 =$$
$$(a^4)\ (a^4)\ (a^4) =$$
$$(a \bullet a \bullet a \bullet a) \bullet (a \bullet a \bullet a \bullet a) \bullet (a \bullet a \bullet a \bullet a) =$$
$$a \bullet a \bullet a \bullet a \bullet a \bullet a \bullet a \bullet a \bullet a \bullet a \bullet a \bullet a =$$
$$a^{12}$$

Again, to clean that up:

$$(a^4)^3 = a^{(4x3)} = a^{12}$$

There are many other properties of exponents. However, those three are the basic ones that students will need to master in the eighth grade. Some teachers may choose to explore more exponent properties with their class but most students won't see any more depth on the topic until their formal algebra class in high school.

Square Roots and Cube Roots

The basic idea of solving an equation is to "undo" the operations happening to the variable. If a variable is squared or cubed in an equation, then a square root or a cube root needs to be taken in order to isolate that variable, that's all.

$$x^2 = 49$$
$$\sqrt{x^2} = \sqrt{49}$$
$$\pm 7$$
$$x = \pm 7$$

It is important to include –7 as well as 7 because both of those values are solutions to the original equation. They both become 49 when squared.

Finding a cube root can be a little trickier. Most calculators don't have a cube root button. Examples of these in class almost always work out nicely, so just guess and check your way to an answer.

$$x^3 = -27$$

$\sqrt[3]{x^3} = \sqrt[3]{-27}$ (The little 3 above and to the left of the root symbol lets the reader know that it is a cube root and not a square root.)

x = the cube root of –27. But what is that? Well, it's negative. So the answer certainly isn't positive. $2 \times 2 \times 2 = 8$, while $3 \times 3 \times 3 = 27$. Excellent. $(-3)(-3)(-3) = -27$, so the answer must be $x = -3$.

Scientific Notation

There are about seven billion people walking the Earth right now. Obviously, there are not *exactly* 7,000,000,000 people. Maybe there are 7,250,147,828;

maybe 7,284,935,201. We round to the nearest billion because the exact number is not overly important.

Numbers are made up of digits and place value. Imagine this: You're on a game show where you see two partially-covered checks, each with your name on them. Check #1 starts with $8, but you cannot see anything after that. Check #2 starts with $3, but again, you can see no more.

Which one do you choose? There is no guarantee that the check starting with the 8 is worth more. Maybe it is, maybe not. $800 versus $30? $8 versus $30,000? It is clear that place value is much more significant to a value than the digits.

Let's take that one step further. You choose the check with the 3. Nice job! The check is worth $3,582,917! First thing you do is call all of your friends to say that you just won "three and a half million bucks!" Wait—do you see what you just did? You ignored $82,917! That amount was not as significant to the story as the first two digits. All you really needed to do was report the information about the largest place values.

This is essentially what scientific notation does. It allows someone to represent a really big number (like 345,000,000,000,000) or a really small decimal (like 0.0000000000007) without having to write all of those pesky zeros every time.

Scientific notation always fits in the following template:

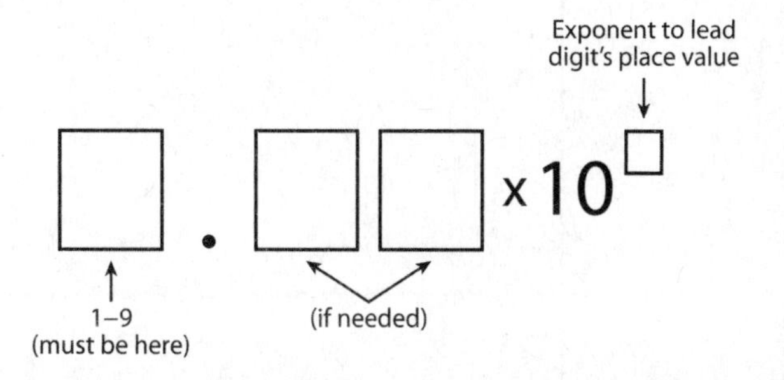

How do I convert 345,000,000,000,000 into scientific notation?

Grab the 345, as they are the digits significant to this number. Put your decimal point after the first digit. 3.45 needs to be multiplied by 100,000,000,000,000 (or 10^{14}) to represent its true place value. 3.45×10^{14} is 345,000,000,000,000 in scientific notation.

Recall what negative exponents do? They represent repeated division. In order to convert a decimal into scientific notation, you would repeat the process but use a negative exponent instead.

Convert 0.00054 into scientific notation.

The 5 and the 4 are the significant figures. Place the decimal point after the 5.

5.4 needs to be divided by ten four times in order to return to the original 0.00054. Therefore, you need to use the exponent −4 on the 10.

$0.00054 = 5.4 \times 10^{-4}$

Graphing Proportional Relationships

Meet babysitter Erin. She makes $10/hour to watch two children. If she was to make a graph comparing how much she makes in terms of how long she works, her points would start as follows:

(0, 0); (1, 10); (2, 20); (3, 30); (4, 40) and so on

Graphically, those points would look like this:

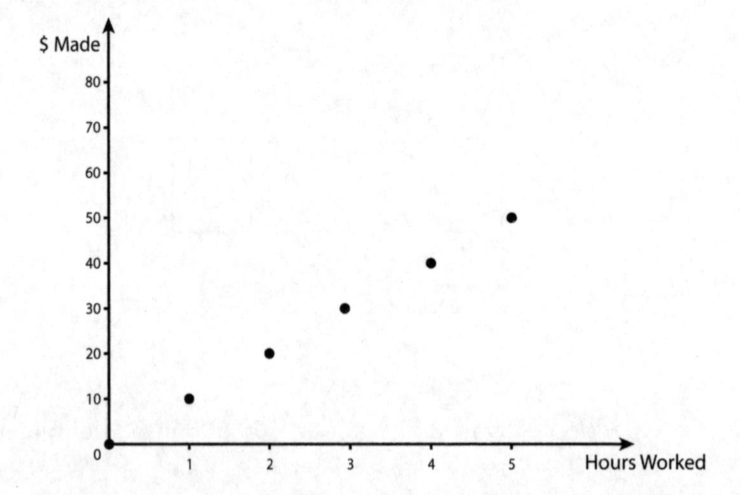

From looking at the graph, you can see that with each hour added, ten more dollars are amassed. The relationship between her pay and her time worked is a *linear function* because the graph forms a straight line. If Erin had some competition, comparing the rates of change between two different graphs would tell who the more expensive babysitter is.

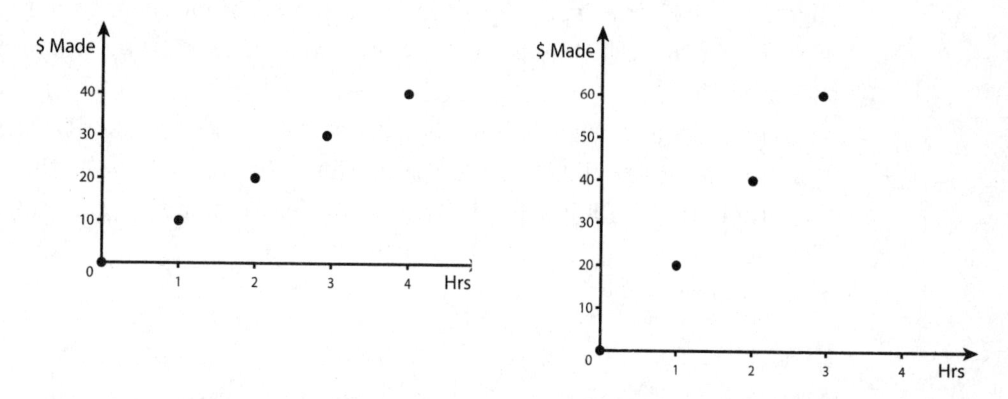

The rate of change is an extremely important concept. It can be expressed as a fraction, seen in a table of values, or found in a graph. It marks the beginning of bridging the English language and equations and graphs.

It is the rate of change that produces the proportion. Erin's $10/hour could create $80 in eight hours, or $120 in twelve hours. These ratios are proportional because they both can simplify back to the same unit rate of $10 per 1 hour.

$$\frac{\$80}{8\ hours} = \frac{\$120}{12\ hours} = \frac{\$10}{1\ hour}$$

What Is Slope?

Slope is huge. There are few things in math that get to show up over and over and over again with new and improved uses each step along the way, as slope does. Slope is even an integral part of calculus. (Pardon the pun!)

Slope has many different ways that it can be expressed. In this chapter, we will touch upon the major ones. However, most concisely, slope is the rate of change in a graph. In other words, "How much more output are you getting every time you add one to the input?"

Recall Erin's babysitting graph. For each hour she works, she makes ten more dollars. The slope of her graph is 10. Oh, and this is as good of a time as any to let you know that the variable m is most commonly used to represent slope. (Why m? Great question. Here's a great answer: *Monter* is French for "to climb.")

The m is used when talking about linear functions generally. The equation $y = mx + b$ is called slope-intercept form, because the m is the slope of the graph and the b is where the graph intersects (intercepts) the y-axis. The idea behind that is that b becomes the entire output when $x = 0$, because m is rendered useless.

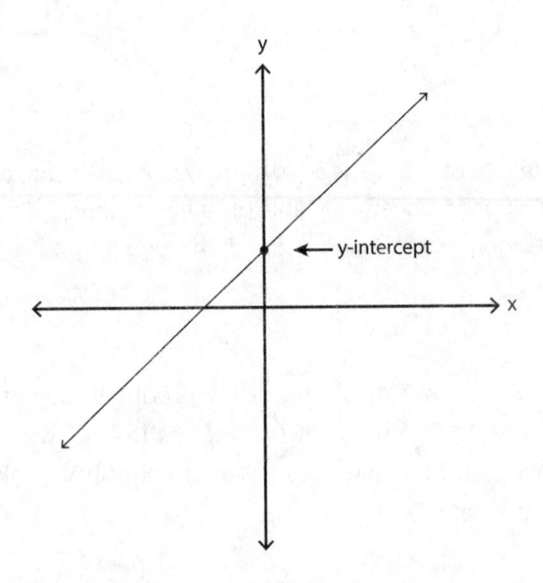

The slope of a linear function is constant throughout the entire graph. The rate of change never increases or decreases. This is what produces the straight line. Other functions, such as $y = x^2$, have curvature to them. The slope is not constant. Examples of these functions are called *nonlinear*.

Sometimes it is difficult to find what the slope is by looking for the increase in the output for one single increase in the input. Cue our next definition of slope!

$$\text{Slope} = \frac{rise}{run}$$

This new rendition of slope holds true in the first example, but it wasn't necessary to use. From one point on the graph, we "rise" 10 and "run" 1 to get to another point.

The term "rise" clearly means "up"—unless, of course, the value is negative. However, "run" *could* mean to the left or to the right. Remember: A positive run is to the right, and a negative run is to the left. If you fear you'll forget, just recall that the *x*-axis increases to the right.

You can find the slope of any straight line by knowing two of its points, either on a Cartesian plane or as written ordered pairs. If you have the graph of the line, find two ordered pairs (preferably ones easy to work with). Starting at one point and moving to the other, how much do you have to rise and how much do you have to run? Throw those numbers into the "rise over run" fraction, simplify if possible, and boom! You've found the slope. For example:

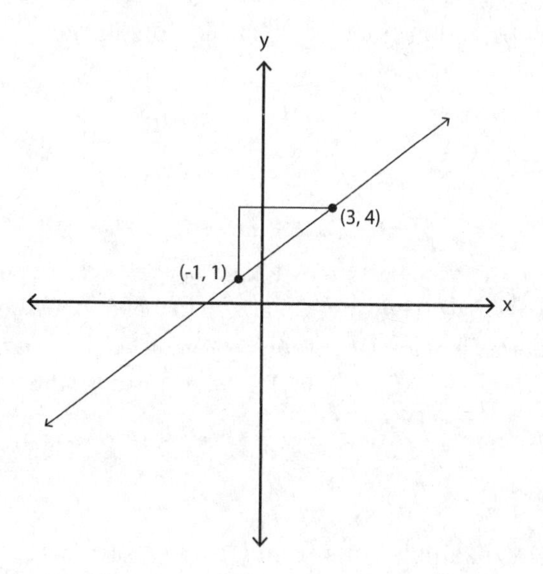

Moving from the bottom-left point to the upper-right point, you rise 3 and run 4. Therefore, the slope of the line is $\frac{3}{4}$.

Only two points are needed when finding slope, because two points define a straight line. Knowing only one point is not enough—you could spin a straight line on that point like a propeller. The moment a second point is added, the propeller is locked into place and cannot spin anymore.

What if you don't have the graph? What if you only have two written ordered pairs? Well then, you are going to need one more definition of slope. Approach with caution, it's going to scare you at first; but, like most math thus far, it is very manageable.

$$m = \frac{\textit{change in } y}{\textit{change in } x} = \frac{y_1 - y_2}{x_1 - x_2}$$

The numbers in the bottom right-hand corners are called subscripts. They do not have value in the problem. They are only there to help delineate between the first y and the second y, as well as the first x and the second x.

Let's look at an example. Say you wanted to find the slope of the line that passes through (3, 5) and (9, –3). You create what could be called the "subtraction fraction." That is, to subtract the y-values on top and the x-values on the bottom.

$$m = \frac{\boxed{5} - \boxed{-3}}{\boxed{3} - \boxed{9}}$$

Two things to be extremely careful with when calculating slope: 1. When working with negatives, it is easy to make a mistake and lose the – of the subtraction problem itself. 2. You can choose either ordered pair to start with, but you must go in the same direction when forming your subtraction problems for *x* and for *y*.

This example runs from left to right when selecting the values to plug into the fraction.

$$m = \frac{5 - -3}{3 - 9} = \frac{8}{-6} = \frac{-4}{3}$$

You now know that if you plotted (3, 5) and (9, −3) on a plane and connected them with a straight line, that the slope of that line would be $\frac{-4}{3}$. For every increase in x (shift to the right once), the output would decrease by $\frac{-1}{3}$ (drop down one and a third). Or, for every increase of three in x, the output will decrease by four.

Solving Linear Equations

There was discussion earlier about how important slope is throughout the continuum of math courses. It pales in comparison to the importance of solving equations. A lot of work has been done in the wonderful world of equation solving thus far. And to be honest, there isn't too much more new information on the subject at hand thrown at students in the eighth grade. The new stuff really isn't new stuff—it's just that some peculiar special cases are considered. More on that in a second.

Almost any linear equation can be solved using the following four-step process:

1. Do any and all distribution
2. Combine like terms within each expression
3. Eliminate the lesser amount of the variable from one side
4. Unwrap the variable

In Step 3, it mentions eliminating the lesser amount of the variable. Honestly, you could eliminate the variable from either side of the equation. However, if you always eliminate the lesser amount, it will always leave your equation with a positive amount of the variable, leaving you with one less thing to think about.

Let's jump right into one of those special cases mentioned earlier to show you the new material students will see at this stage. You'll see a slight hiccup in the four-step process and then, well . . . you'll see.

$4(3x-5)+3x=8+3(5x-8)$	Wow. This is quite an equation that needs to be solved!
$12x-20+3x=8+15x-24$	Start by doing any and all distribution.
$15x-20=15x-16$	Combine like terms within each expression.
$-20=-16$	There isn't a lesser amount of the variable. It's a tie. So subtract that amount of the variable from both sides. But wait a minute . . .

$-20=-16$? It doesn't matter *what* x equals, -20 is *never* going to equal -16. Therefore, write "No solution" as an answer.

If the problem works out such that the variable disappears but you end up with a true statement (like $8=8$ or $-11=-11$), then the original equation is *always* true and you would write "$x=$any number" as your final answer.

Solving Systems of Equations

A system of equations is when you have more than one equation in the same problem. All the same variables equal each other. That is to say that if $x=8$ in one equation, then $x=8$ everywhere else in the entire system.

If given the equation $y=3x-2$, you don't know what the variables equal. If $x=5$, then $y=13$. If $y=25$, then $x=9$. But with only $y=3x-2$, you cannot be certain what the values of the variables are.

To solve a system of equations means to find the value of each and every variable in the system.

You need as many distinct equations as there are variables in a given system in order to be able to solve that system of equations. By distinct equations, I mean that one equation cannot simply be a "multiple" of another. Such is the case in $2x+3y=10$ and $20x+30y=100$. If you have two variables in an equation, you need two distinct equations. Three variables? Three equations. You get the idea.

Remember Erin? The babysitter? She has a competitor, Bridget. Bridget says that she'll work for only *eight* dollars an hour, but she needs ten dollars in gas money to come over. You now have two equations. Hey, you have a system of equations!

We'll let h = number of hours worked, and C = cost of the babysitter. Erin's equation is simply going to be $C = 10h$. Bridget's is slightly more interesting. Hers is $C = 8h + 10$ since the moment she shows up ($h = 0$), she charges ten dollars, followed by her $8 rate of charge kicking in.

If you are curious as to what amount of hours worked results in the same costs, you can solve the system of equations. There are three classic ways to do this.

Data Tables

This is really just an organized way of doing guess and check. Additionally, it only works if the solution is made up of "nice numbers." (The following problem works out nicely.)

▼ **BABYSITTING COSTS**

Hours Worked	Erin's Cost	Bridget's Cost
0	0	10
1	10	18
2	20	26
3	30	34
4	40	42
5	50	50
6	60	58

It is clear that a five-hour shift would cost the same regardless of which babysitter chosen. For less than five hours, Erin is the better choice, and for more than five hours, it's better to go with Bridget.

QUESTION

What happens if the solution is a fraction?
Things could get messy. If the answer is a "nice" fraction, you *may* still be able to find the answer, but there's no guarantee. You'd then be better off using a different method.

Another method would be to graph the two equations. The solution to the system is going to be the point of intersection, because it is the one ordered pair that solves both of the equations within the system.

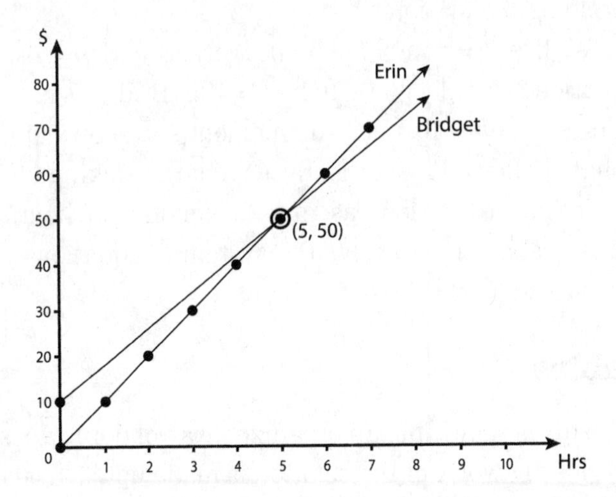

From this method, the solution can be seen as the ordered pair (5, 50). This is actually the solution to the system, because it gives the values of both variables.

If the two graphs are parallel, then no solution exists to the system. There is an infinite amount of solutions to each of the separate equations, but not for the system. No combination of *x* and *y* will possibly satisfy both equations concurrently. In these cases, write "no solution."

The last, and most effective, method of solving a system is to do so algebraically. And there are two ways to do it.

Solving Systems of Equations Using Substitution

In this method, you need an isolated variable. You then substitute (hence the name) its equal expression into the same variable of the other equation. This will leave you with one variable, and you can solve from there:

$C = 10h$

$C = 8h + 10$

$10h = 8h + 10$ (The $10h$ has been substituted in for the 'C.')
$2h = 10$
$h = 5$

And plug the 5 in for h to get $C = 50$.

Solving Systems of Equations Using Elimination

In this method, you manipulate the problem so you have opposite coefficients on one of the variables. Then you add the equations together. This will eliminate one of the variables, leaving you with a one-variable equation to solve.

$C = 10h$
$C = 8h + 10$
$-C = -10h$ (Multiply the first equation through by -1)
$0 = -2h + 10$ (Adding the equations together eliminates the C)
$-10 = -2h$
$5 = h$

At this point, you would do the exact same thing that you did in the substitution process and plug the 5 in for h to get the value of C.

Another name for the process of elimination is the Multiplication-Addition Algorithm. This name fits well, because the process involves *multiplying* equations by numbers to get opposite coefficients, and then *adding* the equations together.

Remember that solving a system of equations means you're finding the value of every variable in the system. The answer should clearly state which value is which. A good solution to the last problem would be $(h, C) = (5, 50)$. But perhaps the best answer would be to write out the answer in words. "After five hours, either girl would make $50."

That was a *lot* of algebra. You deserve a break. Go get yourself a cookie. You can put the book down for a second. I won't be offended.

Practice Problems

Evaluate each.

1. 3^0
2. 5^{-2}
3. $\frac{2}{4^{-3}}$

Simplify each completely.

4. e^4e^6
5. $(3a^2b^3)(ab^5)$
6. $(w^5)^3$
7. $\frac{f^{12}}{f^3}$

Solve for x.

8. $x^2 = 144$
9. $x^2 = 400$
10. $x^3 = 64$
11. $x^3 = -1$

Convert into standard notation.

12. 5.13×10^4
13. 1.08×10^{-7}

Convert into scientific notation.

14. 4,600,000,000
15. 0.000459

16. How much does the electrician make per hour?

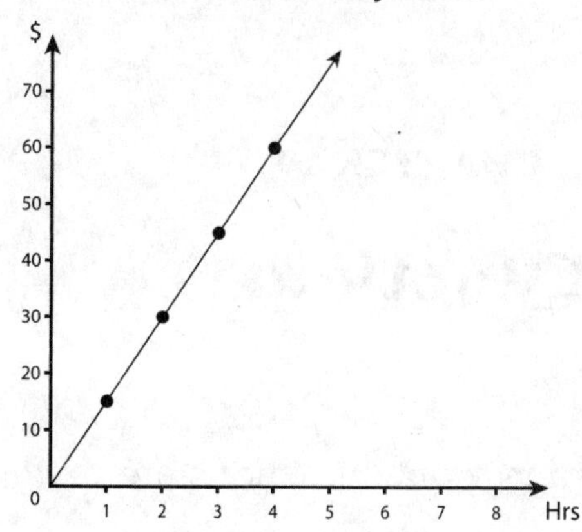

17. What are the slope and the y-intercept of the following line?

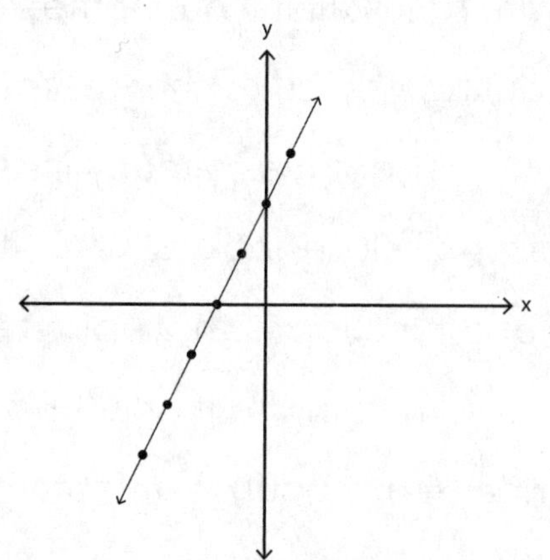

18. What is the slope of the line that passes through (–4, 8) and (2, 4)?
19. Solve for x.

$$3(2x+4)-7=2(4x+1)-x$$

20. Use any method to solve the following system of equations.

$$y=15x+10$$
$$y=5x+40$$

CHAPTER 19

Functions

It's time to put algebra to work. Functions are extremely versatile in their real-life application. They can help an electrician figure out what gauge wire to install in your home's addition, and then aid in figuring out how much to charge. You've seen the word "function" a number of times already in this book, but we really haven't taken a close look at what it actually means. The basic idea of a function is this: You start with a number (an input), you do something to it (the function rule), and you get a new number (an output). We match up each input with its output (written as an ordered pair). And the collection of *all* of the ordered pairs is called a function. It can be represented by an equation, a table of values, or as a graph. If you monkey around with the operations in the function, it changes the appearance of the graph. There is, however, one very important rule with functions—*no matter what input you use, you can only get one output.*

What Should My Child Already Be Able to Do?

Since students will be hopping back and forth from functions to graphs, they should have an understanding of how to plot ordered pairs. They will also see a lot of computation in this unit, so that brings us back to the importance of arithmetic skill. Most of the function rules seen in this unit will be a combination of addition, subtraction, multiplication, and division, with maybe some squaring sprinkled in for fun.

Truth be told, eighth grade will not be the first time that students will have seen functions. As early as the first grade, students will be asked to do such math as:

Add five to each of the following numbers

$$2 + \underline{5} = \_\_\_\_\_$$

$$7 + \underline{5} = \_\_\_\_\_$$

$$0 + \underline{5} = \_\_\_\_\_$$

$$10 + \underline{5} = \_\_\_\_\_$$

Guess what: that's a function rule. The middle-school curriculum upgrade has students not only using more complicated operations, but also comparing different functions. There is a lot of commonality between comparing functions graphing proportions. A lot of what students will explore in this section will be reminiscent of the proportion comparing that they did earlier in a previous unit.

What Is a Function?

Before we define exactly what a function is, we need to recall what a relation is. A relation is a set of ordered pairs. {(1, 2), (5, 7), (5, 9)} is a relation. However, it is not a function. A *function* is a relation where every input used is paired up with exactly one output. Since the 5 gets paired with 7 as well as with 9, it loses its privilege of being called a function.

Although a function cannot have the same input result in multiple outputs, it *is okay* for multiple inputs to result in the same output. For example, the relation {(2, 5); (3, 5); (4, 6)} *is* a function, because none of the inputs are matched with multiple outputs. As an analogy, consider that $4+5=9$ and $3+6=9$. Two different "beginnings" may have the same "end," but no "beginning" had multiple "endings." If $3+6$ had multiple correct answers, math wouldn't "function" very well at all!

There is another way to check to see if a relation is a function. It is called the *vertical-line test*. Start by plotting or graphing the relation. We'll use the previous example here.

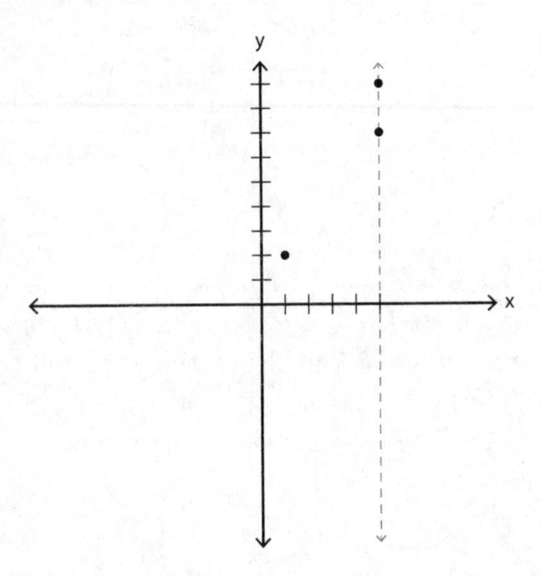

This relation is not a function because two points appear on the same vertical line.

Imagine a vertical line scanning the graph, sweeping from left to right. There isn't a blip on the radar until the vertical line reaches $x=1$. It hits a point at (1, 2). No problem there—one output for the input of 1. However, a short while later, the vertical line passes over $x=5$ and hits *two* outputs. No good. Or, no good if you were hoping for a function.

Function Notation

Writing something in function notation guarantees that the equation is, in fact, a function. There is no ambiguity here. Typically, functions are written with x being the input variable and y being the output variable. In function notation, any variable can be chosen for the input, but then the output is written a little differently. If t is chosen as the input variable, then $f(t)$ is the output. This stands for "the function applied to t," and is read as "f of t."

The *f* in *f*(*t*) is not a variable representing a number. It represents the entire function. It is usually written as a cursive letter to distinguish it from other types of variables. Sometimes, instead of the *f*, you will see a capital letter that stands for something specific. For example, *H*(*t*) may be used to represent the function "Height in terms of time." Again, no number is substituted in for the *H*.

A teacher writes the function $f(x) = 4x - 5$ on the board. She then asks students three questions: Find $f(2)$, $f(-5)$, and solve for x if $f(x) = 23$. To answer the first question, the students are going to plug 2 in wherever they see an x. $f(2) = 4(2) - 5$ and therefore, $f(2) = 3$. Using the same process, $f(-5) = -25$. The last problem is done in reverse fashion. It's asking what value of x will make the function worth 23. So, replace $f(x)$ with 23, and then solve the equation $23 = 4x - 5$ to get $x = 7$.

Define, Evaluate, and Compare Functions

At this point, students have seen function notation and explored graphing equations on the Cartesian plane. Naturally, it would make sense to next combine these two topics together. Students will be asked to graph functions and draw conclusions about what they see. This is typically when an eighth grader will be asked to graph a nonlinear function (an equation that produces a graph that is not a straight line).

If the only operations done to an input variable are a combination of addition, subtraction, multiplication, and division by known numbers, the function will be linear and produce a straight line. Raising the input to an exponent and taking the absolute value of the input are two examples of functions whose graphs will not be a straight line.

All linear functions can be written in the form $f(x) = mx + b$ where m is the slope and b is the y-intercept. The only variation from linear function to linear function will happen when m and b are changed. Changing b simply raises or lowers the entire line to intersect the y-axis at the given value.

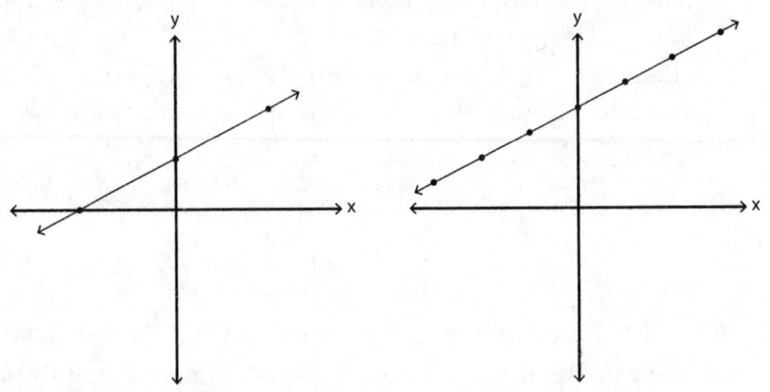

Both graphs have equal values of *m* so they are parallel. Changing the value of *b* raised the line.

Changing m will make the line more or less steep.

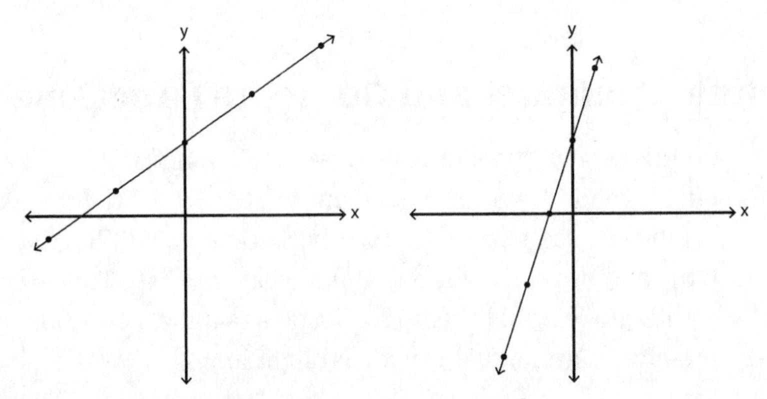

Both lines have the same *y*-intercept. Increasing *m* from $\frac{2}{3}$ to 3 made the line steeper.

A slope of 0 will produce a perfectly horizontal line. Slopes, positive or negative, make their graphs steeper the further they get from zero.

What is the slope of a vertical line?
The slope of a perfectly vertical line is undefined. Yep, it literally does not exist. A vertical line would fail the vertical-line test, meaning that the relation is not a function. Additionally, since slope can be calculated as $\frac{rise}{run}$, and any two points on a vertical line would produce a run of 0, you'd be dividing by zero and that is not allowed!

At this stage of the game, graphs of nonlinear functions are going to be produced using a data table. Students will choose the numbers they want to use as inputs, and evaluate their corresponding outputs by repeatedly applying the function rule.

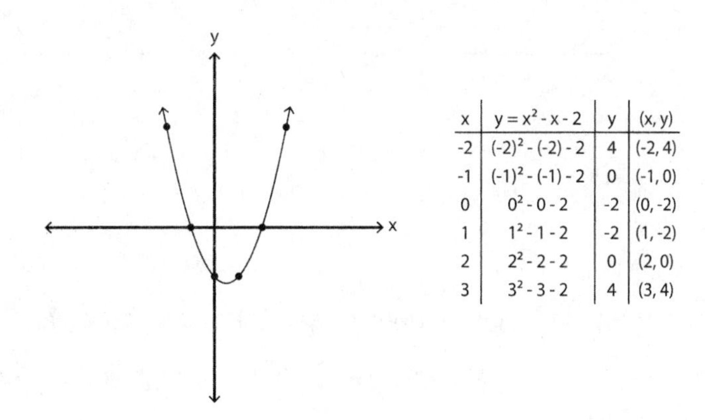

x	$y = x^2 - x - 2$	y	(x, y)
-2	$(-2)^2 - (-2) - 2$	4	(-2, 4)
-1	$(-1)^2 - (-1) - 2$	0	(-1, 0)
0	$0^2 - 0 - 2$	-2	(0, -2)
1	$1^2 - 1 - 2$	-2	(1, -2)
2	$2^2 - 2 - 2$	0	(2, 0)
3	$3^2 - 3 - 2$	4	(3, 4)

The following are three examples of functions that look extremely different.

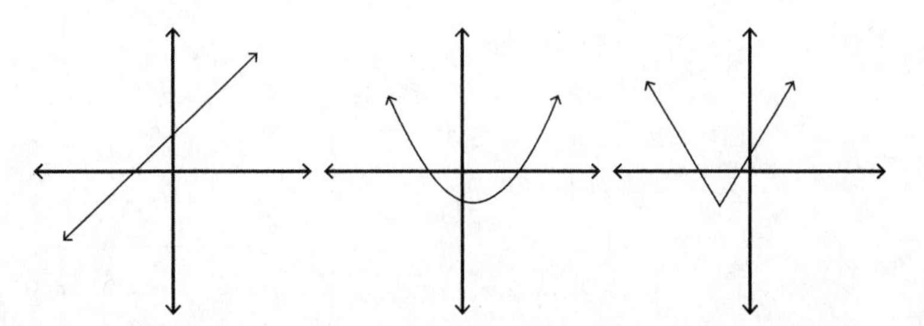

The first graph can be produced with a basic understanding of slope and the y-intercept. However, the second two graphs will most likely be produced from a data table. Functions where the input is being squared produce smooth curves called parabolas. Functions where the absolute value of the input is being taken produce V-shaped graphs. Students will explore those last two types of functions in much greater detail in their formal algebra course. This course usually immediately follows the eighth-grade Common Core class.

Practice Problems

For problems 1–3, decide whether or not the relation is a function.

1. {(1, 4); (3, 5); (3, 7); (5, 10)}
2. 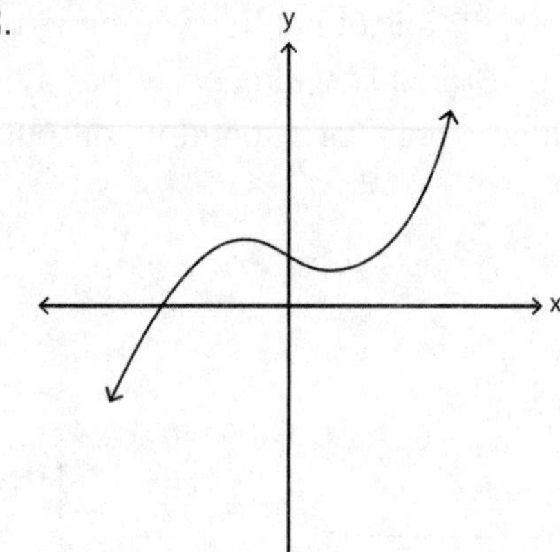

3. 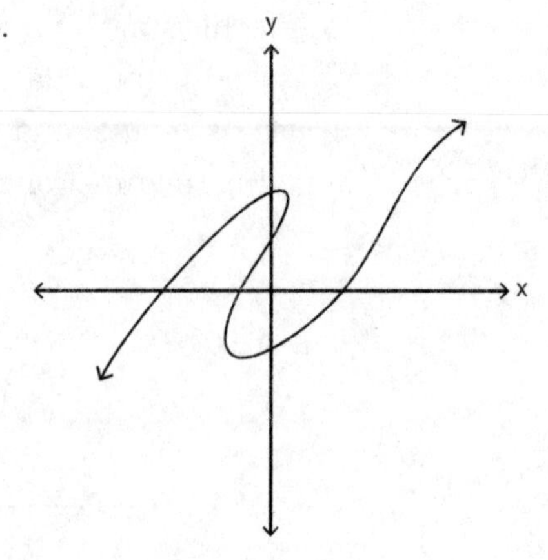

Use the following functions to answer questions 4–7.

$$f(x) = 4x - 5\ ;\ g(x) = 2x^2\ ;\ h(x) = x^2 - 3x + 9$$

4. Find $f(-10)$.
5. Find $g(3)$.
6. Find $h(5)$.
7. If $f(x) = -13$, find x.
8. Complete the data table provided and use that information to graph $f(x) = x^2 - 2x - 2$.

x	y
-1	
0	
1	
2	
3	

CHAPTER 20

Geometry

Geometry is typically one of the favorite sections among eighth graders. Because there are specific formulas to use for each particular situation, students who are good at rote mathematics tend to excel. However, that is not to say that there is no opportunity for creativity throughout the unit. In eighth grade, students make the leap from the second dimension to the third. Those who love to draw will certainly enjoy sketching representations of three-dimensional figures, or "solids." Difficulties in this section usually stem from an inability to visualize three-dimensional objects based on two-dimensional drawings.

What Should My Child Already Be Able to Do?

Students should have a basic understanding of scale factors and their significance in geographic maps. In the eighth grade, students will take what they know of scale models and transfer it onto the Cartesian plane. If your school chooses to present geometry before the functions and graphing unit, you will want to make sure that your student has a solid understanding of how to plot an ordered pair.

How do I plot an ordered pair?
Remember, ordered pairs are written (x, y) and the x-axis is the horizontal axis. So, to plot an ordered pair, start at the intersection of the x and y-axes, move to the side for the x value (left if negative, right if positive) and vertically for the y value (up if positive, down if negative).

Although students in seventh grade work with two-dimensional shapes as well as three-dimensional solids, the major focus is on shapes. By the time students reach the eighth grade, they should be well versed in finding the area of rectangles, triangles, and circles. The process of finding area is embedded within most of the formulas for calculating the volumes of the solids the students will see in the Common Core State Standards.

Basic Area Formulas

Rectangles (and squares): Area = length × width

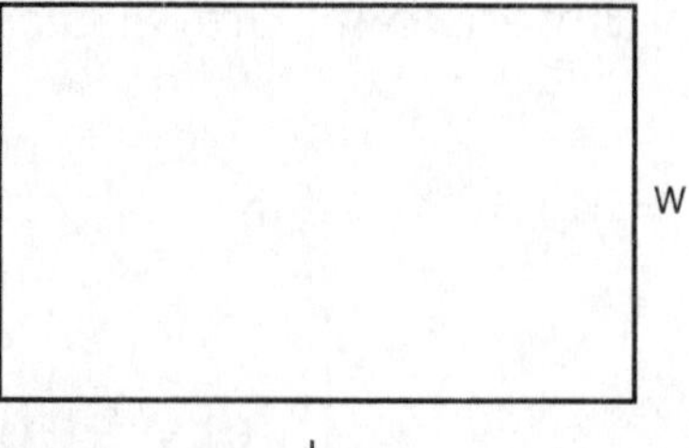

Triangles: Area = $\frac{1}{2}$ of the base × height (where the height and the base meet at a right angle)

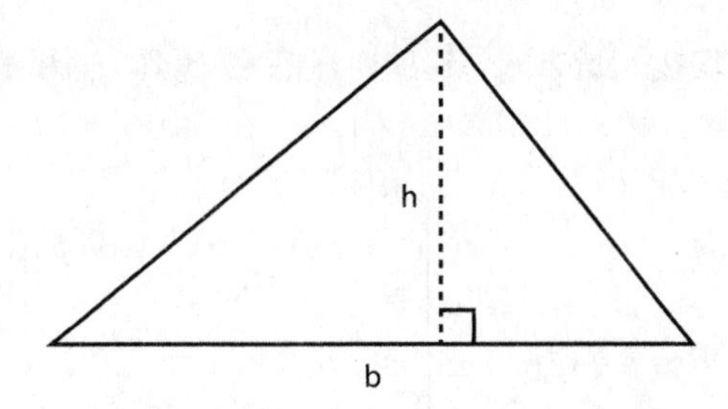

Circles: Area $= \pi \times r^2$ (where r stands for radius, or half of the distance across the circle)

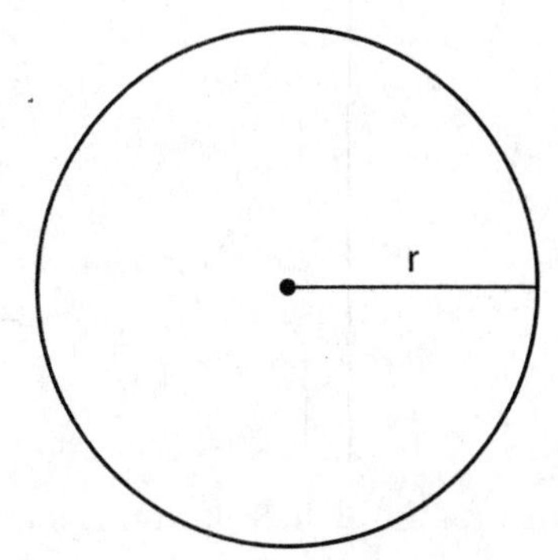

π is the Greek letter pi. Mathematically, it is used to represent the ratio between a circle's circumference (the distance around the circle) and its diameter (the distance across the circle.) Pi is a decimal that never ends, so most people simply use 3.14 as an approximation.

Also covered in eighth-grade geometry is the Pythagorean Theorem. It is used to find missing side lengths in right triangles. Students will be expected to know that a right triangle is any triangle that has one angle that measures exactly 90 degrees.

What Is the Difference Between Similarity and Congruence?

Congruent means "equal" in the world of geometry. Simply put, numbers can be equal, while shapes can be congruent. It's all in the wording. Two angles are considered *congruent* when their angle measurements are *equal*. When we talk about numbers and values, two of the same are equal. But when we are talking about angles, line segments, two-dimensional shapes, three-dimensional solids, etc., two of the exact thing are called congruent. You get the idea.

Similarity however, is a bit different.

Picture a balloon being inflated. Once its general shape is established, the balloon gets bigger but it stays the same shape.

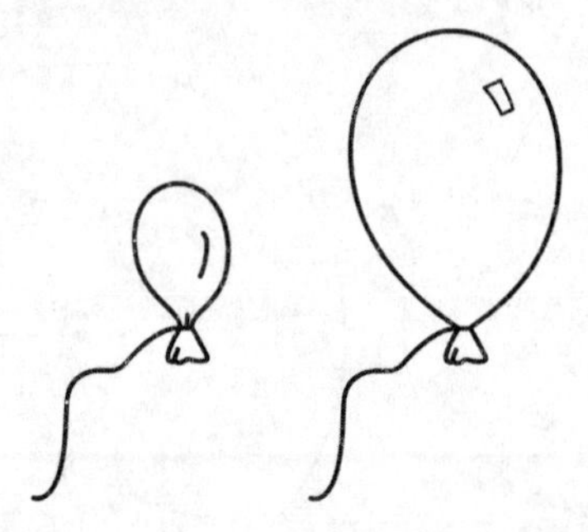

Now picture a triangle. Imagine it somehow being inflated. Each of the three angles will stay the same, while the side lengths will all grow at the same rate.

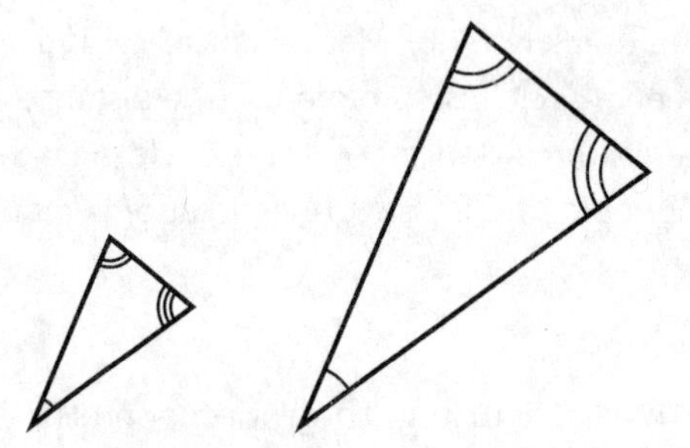

The triangle you started with and the triangle you finish with are *similar*. While any geometric elements (like line segments and angles) can be considered as possibly congruent to others, only closed shapes (like rectangles, circles, and stars) can be similar to others.

In similar shapes, corresponding angles are congruent, and corresponding lengths are proportional.

Students may make the mistake that angle measures get scaled up as well as side lengths in similar figures. However, one needs to look no further than two different-sized squares to confirm that the angles must stay congruent. Any square, regardless of size, has exactly four angles of exactly 90 degrees each. Additionally, because of their specific shapes, any two circles are similar to one another.

Transformations: Flip, Turn, Slide, and Dilate

With the exception of dilations, *transformations* are nothing more than actions taken to move an image from one place to another. Flip, turn, and slide are more common terms for reflection, rotation, and translation, respectively. Since all that is happening is a movement of a drawing, these three transformations preserve congruence. We will explore each of these first three in a little more detail before we get to dilations.

Reflections

Much like how you see your reflection every time you look in the mirror, a reflection in math creates a mirror image of a design over a designated line. This line is called the *axis of reflection*.

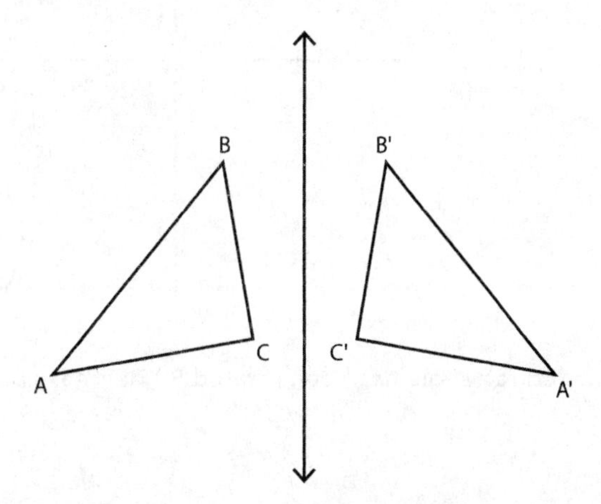

Triangle ABC has been reflected over the line of reflection (*l*) to become triangle A'B'C'.

The new point that an original point gets moved to is called the *prime* of the old point. Prime points are denoted with a prime symbol, which looks like a straight apostrophe. Notice on the previous page how point A matches up with A', B with B', and so on.

Additionally, notice that the direct distances of any original point and its prime to the axis of reflection will always be equal.

Rotations

A simple line is all that is needed to know where to reflect an image, but a bit more specificity is needed in order to perform a rotation. You first need to know the *point of rotation*. It will serve as the one set point around which the image will be spun. (Picture a poster being tacked onto a bulletin board with only one tack. The single tack serves as the point of rotation and the poster can be spun around it.) You also need to know how many degrees to spin the image, and lastly, you need to know which direction to go—clockwise or counterclockwise.

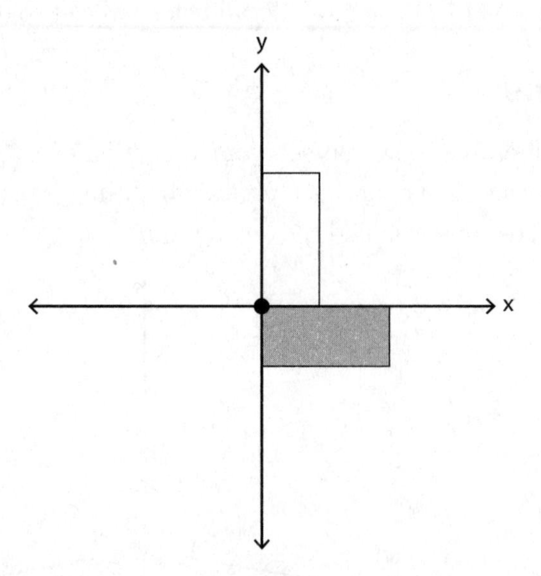

The shaded rectangle has been rotated 90 degrees counterclockwise around (0, 0).

Rotating 180 degrees is the only time that direction is not necessary.

A piece of tracing paper can be very helpful when rotating an image. Trace the original shape on the paper, press your pencil tip down on the point of rotation, and then spin the paper accordingly.

Translations

Translations occur from any combination of sliding an image up, down, left, and/or right. Translations not only preserve congruence, they also preserve orientation, or the direction the shape is viewed.

"Translation" sounds a lot like "transformation." Do not get them confused. A translation is just one example of a transformation.

A *translation* needs to be explained with not only a direction to move, but also a distance. Any movement can be broken down into components; that is, the amount to move horizontally and the amount to move vertically. You may see a translation being described by a strange-looking ordered pair. For example, $(x+6, y-4)$ means that the image should be shifted six units to the right (adding to the x-values of each ordered pair) and down four units (subtracting from the y-values of each ordered pair).

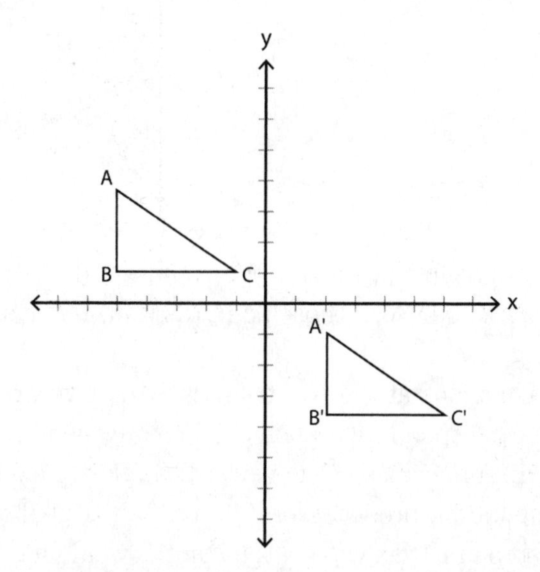

Triangle ABC has been translated by $(x+6, y-4)$ to become triangle A′B′C′.

Dilations

When you experience a change in light, the pupils of your eyes will dilate to either allow more or less of that light into your eyes. The shape of your pupils doesn't change—only the size. *Dilations* do the same thing to an image.

All that is needed to perform a dilation is the scale factor. That is, the amount to increase or decrease the lengths of the image. Try this:

Plot the points (2, 4), (3, 1) and (0, –1) on the following plane. Connect those points into a triangle.

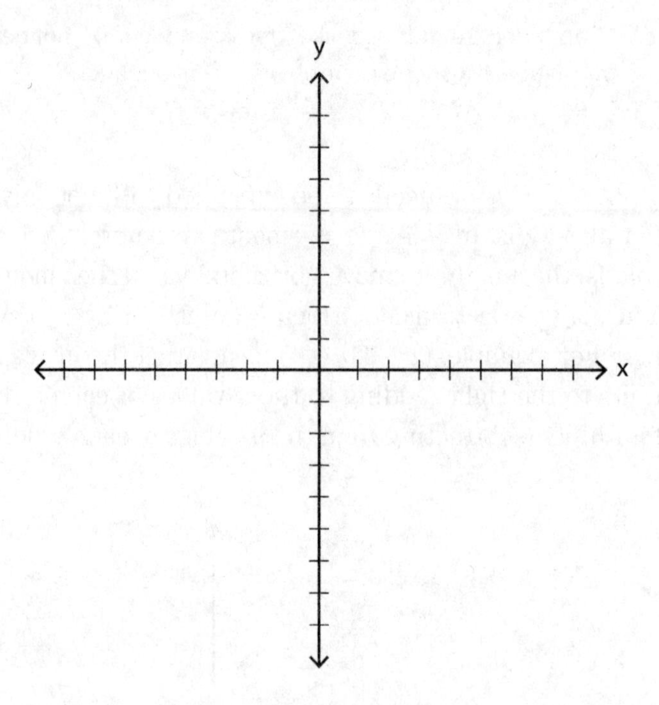

If you double every number in those ordered pairs, you get (4, 8), (6, 2), and (0, –2). Now plot those points as well, and again connect the points into a triangle.

Take a look at the two triangles you have drawn. They are similar! All of the corresponding angles will be congruent. And, maybe more interestingly, all three sides of the newer triangle are exactly twice as long as their corresponding predecessors. If you had tripled the values in the original ordered pairs, then yep . . . the lengths would have tripled as well.

The Pythagorean Theorem

This book won't delve into the history of the Pythagorean Theorem, but it is sufficient to say that it has been in use for a long, long time. Its basic idea is unbelievably simple and elegant. It is literally as easy as "abc."

The Pythagorean Theorem states that "given any right triangle, the square of the length of the longest side will equal the sum of the squares of the two shorter sides' lengths."

Wait a minute! That didn't sound simple at all! So let's step back for a second and take a closer look at a right triangle.

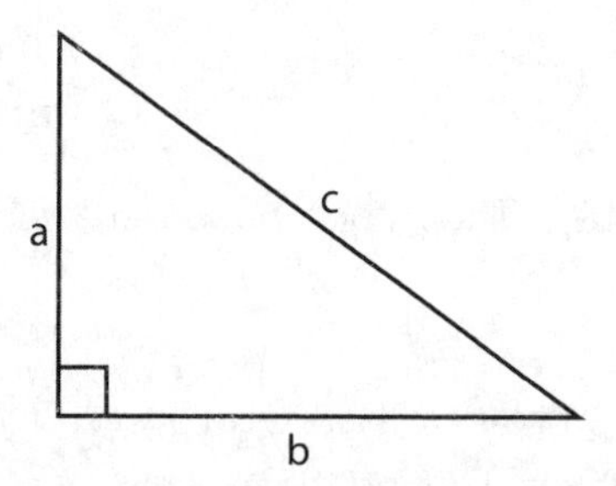

A right triangle with side lengths *a*, *b*, and *c*.

The side across the triangle from the right angle will always be the longest side. We call this side the *hypotenuse*. It is typically labeled *c*. The other two sides are each called *legs*.

Although there is no significance in the labeling of the legs, the shorter of the two legs is typically labeled *a*, and the longer of the two legs is typically *b*.

With the three lengths each having a variable assigned to them, you can see the true simplicity of the Pythagorean Theorem:

$$a^2+b^2=c^2$$

If you know two of the three lengths in a right triangle, you can use the Pythagorean Theorem to find the third.

For example, if a right triangle's two shorter sides have lengths 6 inches and 8 inches:

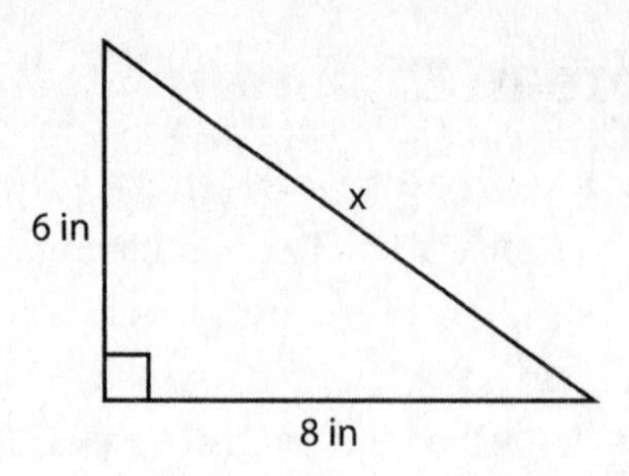

$a^2 + b^2 = c^2$

$6^2 + 8^2 = c^2$ Substitute the known lengths into the formula.

$36 + 64 = c^2$ Square the two known lengths.

$100 = c^2$ Simplify.

$10 = c$ Find the square root of both sides.

The length of the hypotenuse is 10 inches.

Not every problem will work out as nicely as the previous example. If the result is irrational, round the decimal.

If the missing length is a leg, the process changes slightly. For example, find the missing length:

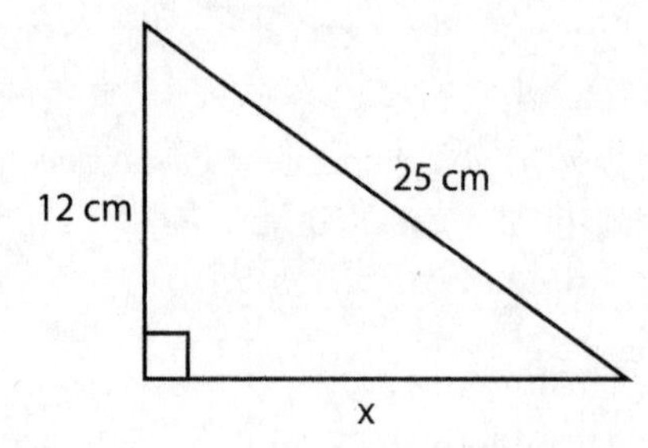

$a^2 + b^2 = c^2$

$12^2 + b^2 = 25^2$

$144 + b^2 = 625$

$b^2 = 481$ In this problem, subtract instead of add.

$b \approx 21.93$ Since 481 is not a perfect square number, round the decimal.

The missing length is about 21.93 cm.

If a value gets rounded, the answer becomes approximate and the symbol $\approx$ is used, instead of the usual $=$.

Volume of Cones, Cylinders, and Spheres

The good news is that volume formulas are very rote and reliable. Once a formula is memorized, the same formula can be used every single time a volume needs to be found.

The bad news is that it can sometimes be very tricky to visualize a three-dimensional solid from a two-dimensional drawing. Teachers will often have tangible models of the solids being discussed for students to hold and explore. This can be extremely helpful.

When finding area, the unit of measure is always in units *squared*, since area is a two-dimensional metric (e.g., in^2, cm^2, mi^2). However, volume is a three-dimensional metric so be sure to use *cubic* units (e.g., in^3, cm^3, mi^3)!

Prisms

Finding the volume of a prism is a very straightforward concept. But in order for things to run smoothly, you'll need to explore what makes a prism a prism.

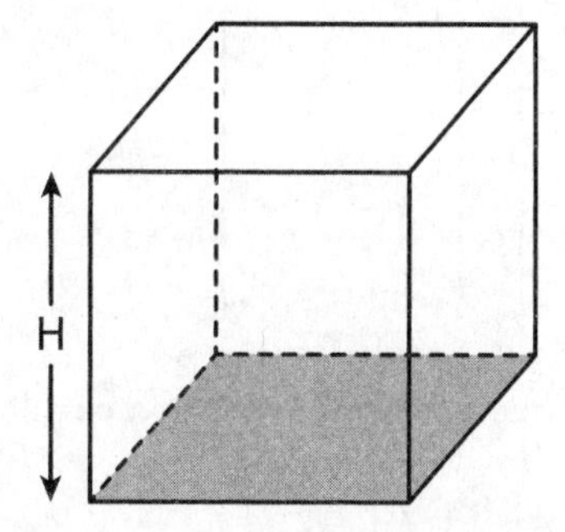

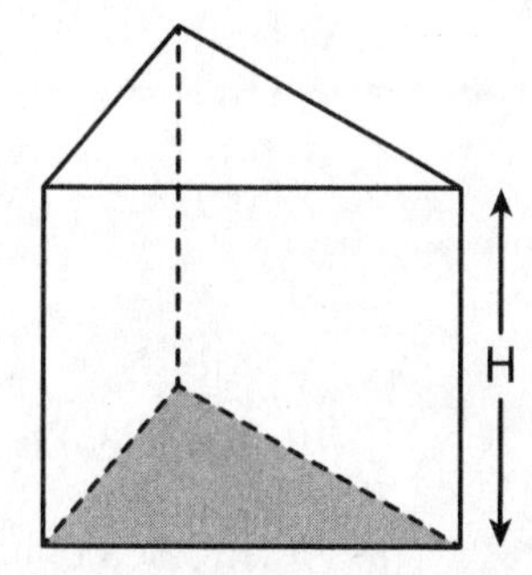

If these prisms were filled with water, the size and shape of their surface would never change.

A *prism* is created when a polygon is maintained throughout an entire height. You can picture clay being pushed through an opening. The clay being extruded is going to get longer and longer, but the cross section will always be the same shape. In the previous pictures, the rectangle and triangle will be the cross section at any instance of the solid being "sliced."

The length of extrusion is called the *height* of the prism. The height of a prism is the distance between the congruent bases, no matter how the solid is held—even if it is lying down. Consider this: A six-foot-tall person is still considered to be six feet tall even if he is seated or lying down napping.

The method for finding the volume of any prism is the same. Find the area of the base and multiply by the height of the prism.

If the base of a prism is a triangle, it will have its own base and height. These are not the same as the base and height of the prism. The area of the triangle must be found first before multiplying by the height of the prism in order to find the volume.

Cylinders

Cylinders aren't *technically* prisms, but they have all the same characteristics with one exception—their base is always a circle. Since the formula for finding the area of a triangle is always the same, the volume formula for cylinders will always be the same.

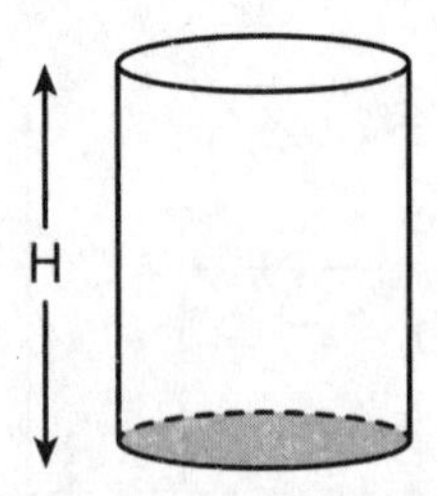

A cylinder is a lot like a prism but it has a circular base.

The volume of a cylinder $= \pi r^2 \bullet H$

Cones and Pyramids

Amazingly, the volume of a cone is exactly one-third of a cylinder with the same base and height. And analogously, the volume of a pyramid is exactly one third of a prism with the same base and height. This is fairly nice in the world of volume formulas.

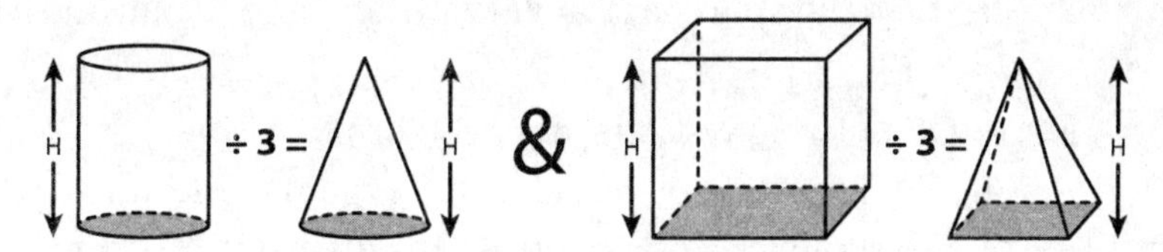

Volume of a cone: $\pi r^2 \bullet H/3$ Volume of a pyramid: (Area of base) $\bullet H/3$

Spheres

Spheres are a bit unique. They have their own volume formula. But again, all spheres have the same basic design. Spheres will only differ by their size. Therefore, the same formula can be used to find the volume of a sphere.

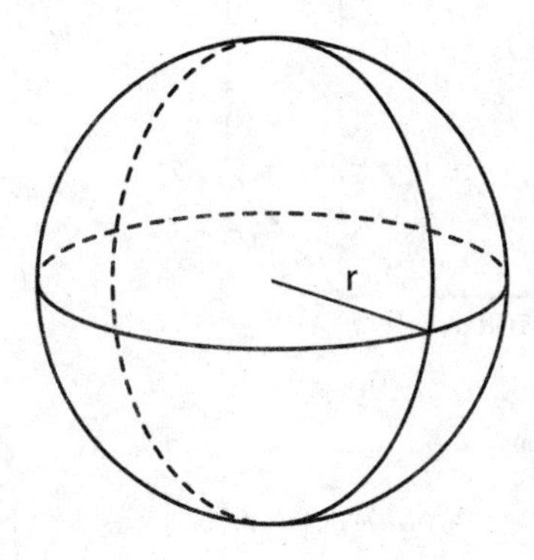

Volume of a sphere: $\frac{4}{3} \bullet \pi r^3$

The measurement needed to find the volume of a sphere is the radius. If the diameter is given (the distance across the entire sphere), divide it by two to get the radius.

Practice Problems

1. Find the area of a rectangle with length 6 cm and width 10 cm.
2. Find the area of a rectangle with height 7 inches and base of 12 inches.
3. Find the area of a triangle with a base of 10 mm and a height of 16 mm.

For the next two questions, use 3.14 for pi.

4. Find the area of a circle with a radius of 12 ft.
5. Find the area of a circle with a diameter of 14 cm.

Find the volume of each of the following solids.

6.

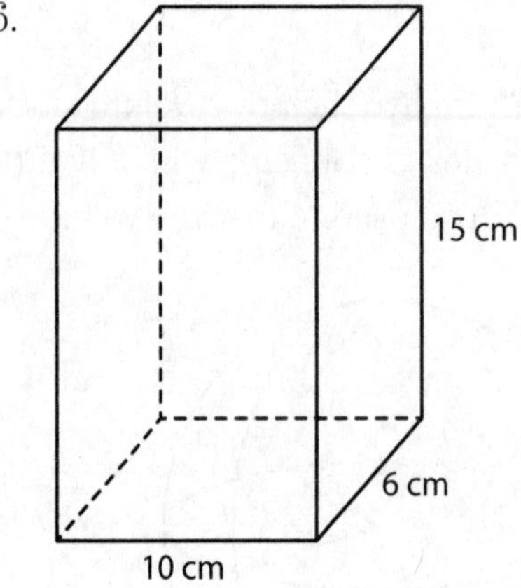

7.

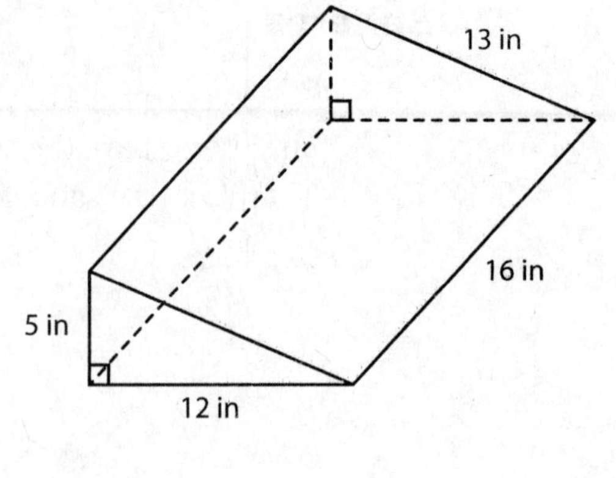

8.

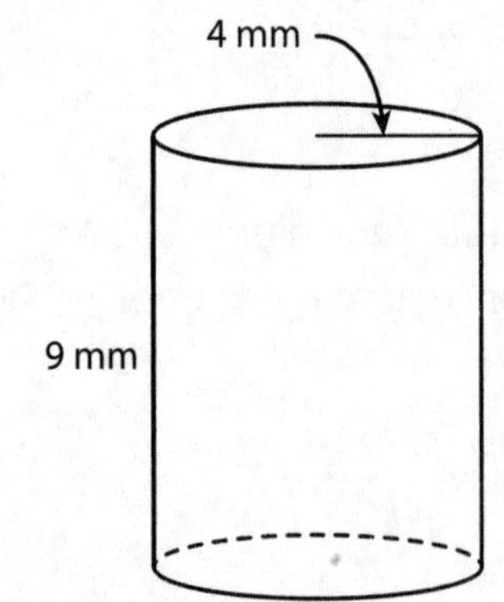

9.

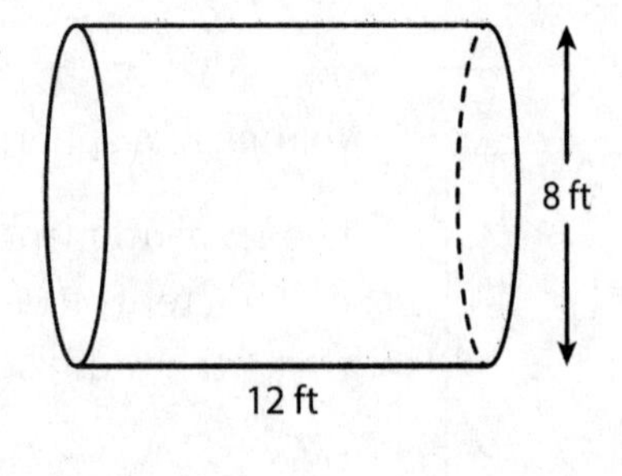

10.

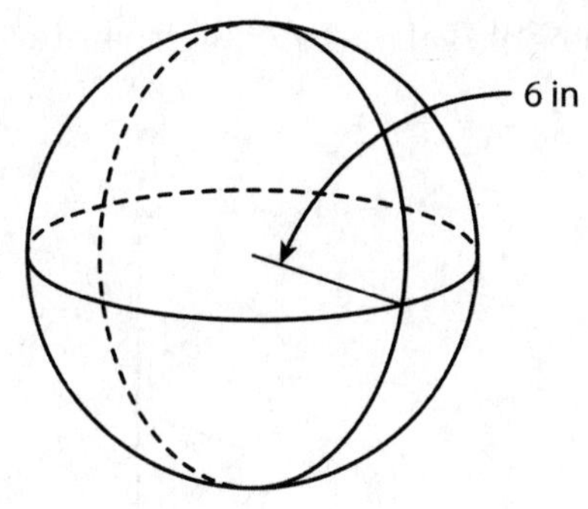

11.

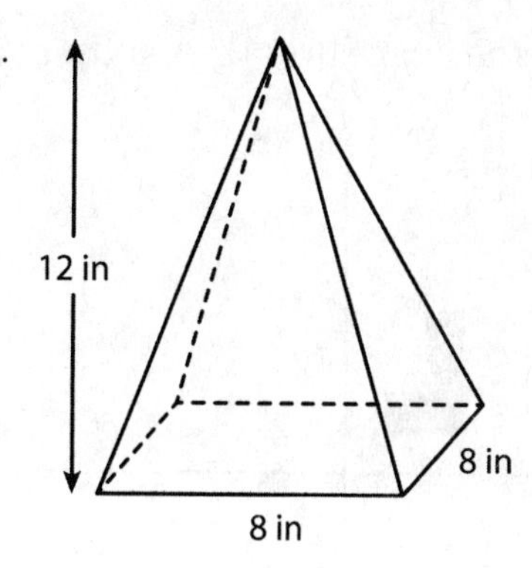

12.

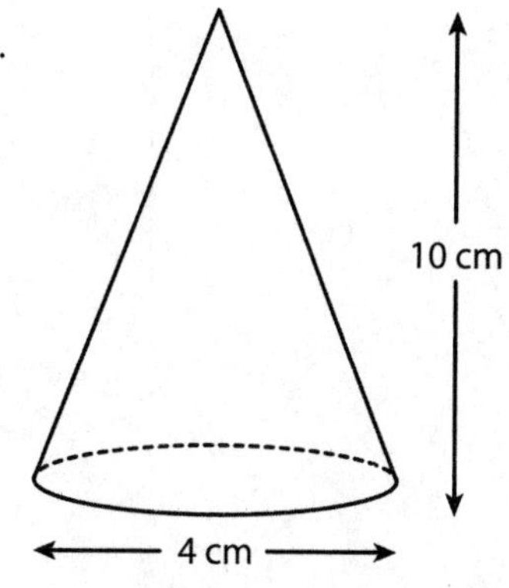

13. Find the length of the hypotenuse of a right triangle whose legs have a length of 8 inches and 15 inches.
14. Find the length of the third side of a right triangle whose longest side is 50 mm long and whose shortest side is 30 mm long.
15. Translate ΔABC by $(x+3, y-2)$.

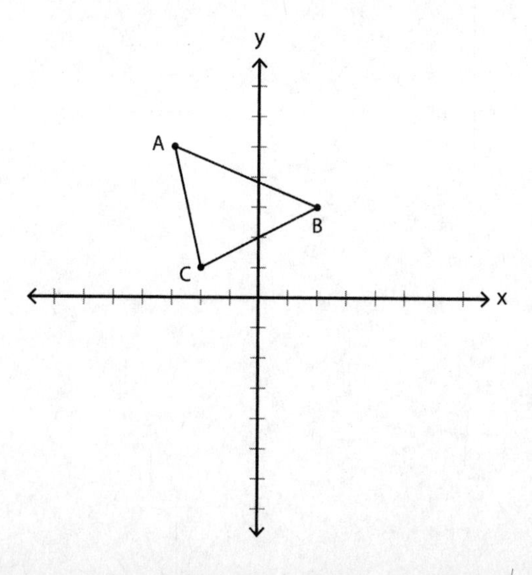

16. Reflect the image over the y-axis.

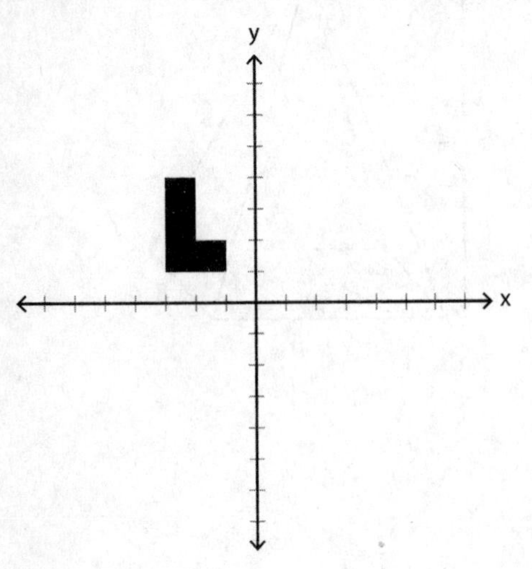

17. Rotate ΔDEF 90° clockwise about the origin.

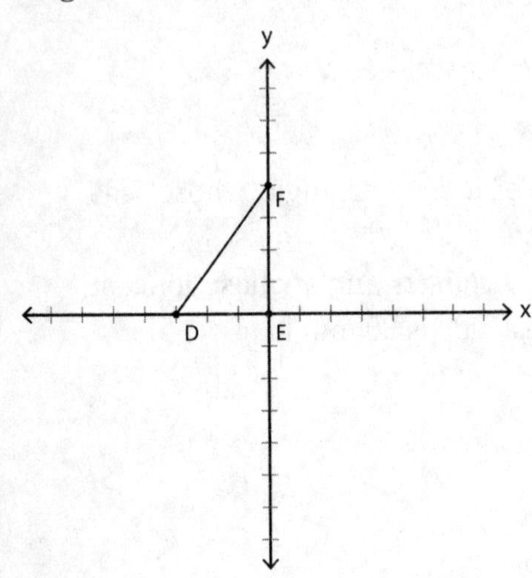

18. Dilate the quadrilateral by a scale factor of 2.

CHAPTER 21

Statistics and Probability

Cause and effect; it's an important lesson for any middle-school student to learn in general. Aside from the life lessons that they absorb along the way, cause and effect is a fairly big deal in the curricula. Social studies teachers discuss how one man's actions have a domino effect on his surroundings. Science teachers hypothesize about what effect may arise from an experiment. In math, we get the ball rolling with basic measurable data within an input-output framework. This is where you'll see one of the strongest correlations between math and science.

What Should My Child Already Be Able to Do?

There is very little probability in the eighth-grade curriculum. Within this stream, most of what students will see this year will be statistical. It will be expected that students will come into this unit having already mastered the basics of calculating probability.

At this point, this book has discussed a huge amount of material based upon functions and graphing equations. Bivariate data is based around the input/output nature of functions. The majority of the correlations discussed in the eighth grade will be linear. Therefore, students will revisit the $y = mx + b$ form early and often.

Are there any other forms of linear equations other than slope-intercept ($y = mx + b$)?
There certainly are! $Ax + By = C$ is called *standard linear form*, and $y - y_1 = m(x - x_1)$ is called *point-slope form*. Both serve their purpose very well when the time is right. However, since the emphasis this year is on the rate of change in linear functions, and slope-intercept is so direct and convenient, students usually don't see the other two forms until their formal algebra course in high school.

Often in math, students learn a process one way, then the setup of the question is spun around and students are asked to work with a problem in reverse fashion. For example, students may be asked to perform a transformation on an image. Then they may be asked to describe a transformation that was already performed. Such is the case with graphing in this unit. Instead of being given an equation to graph, students will be given bivariate data, plot it, and then extrapolate the equation from the graph. Needless to say, students should have a thorough understanding of the roles of m and b on a graph.

Bivariate Data: What Is It?

Bivariate simply means two variables. The variables in this case represent something meaningful. Well, maybe not intensely meaningful to the students, but to *someone*.

In statistics, variables must be quantifiable, usually with some type of unit of measurement. For example, "height" is a perfect variable in statistics because it is measured numerically. Also, it has a choice of a few different well-defined units of measurement (such as centimeters or inches). "Happiness" is not a good variable because it is subjective, non-numerical, and devoid of a unit of measurement.

The frequency (how many) of an item can also be used as a variable in statistics. For example, the fuel economy of a pickup truck is certainly related to the number of patio paving stones loaded into the back.

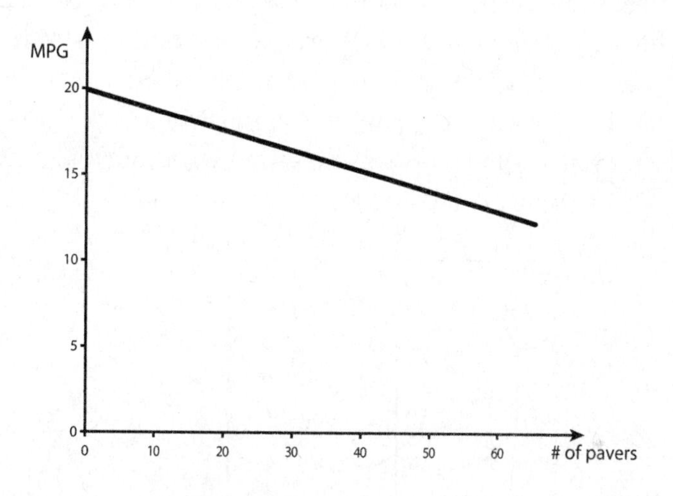

The variables from the previous example have a *negative correlation*. That is to say that as the input variable (the number of paving stones) increases, the output variable (fuel economy) decreases. If the output increases as the input increases, then the two variables have a *positive correlation*. If there is no discernable connection between the variables, then there is *no correlation* between them.

Essentially, this unit mirrors the linear function unit, but assigns meaning to the domain and the range. Calculating and comparing rates of change is an essential piece of the puzzle. If you know the fuel economy of your empty pickup truck and how many miles per gallon you lose for each paving stone you toss in the back, then you can figure out the function "economy in terms of paving stones," or "$E(p)$."

Two-Way Tables

Two-way tables are one of the newest additions to the middle-school curriculum. They are a little intimidating, because they look a little different and they contain so much information in such a small amount of space. However, once you pick up on the patterns of how they work, they are actually quite user-friendly.

Two-way tables only work to compare *either/or* characteristics about things. Examples include male or female when talking about people, day or night when talking about baseball games, and wooden or other when talking about roller coasters. When two options are structured in a rectangular array, a grid of the four possible combinations emerges.

The information can also be displayed in a Venn diagram. To demonstrate the connection between the two, let's explore the following example of a Venn diagram being converted into a two-way table.

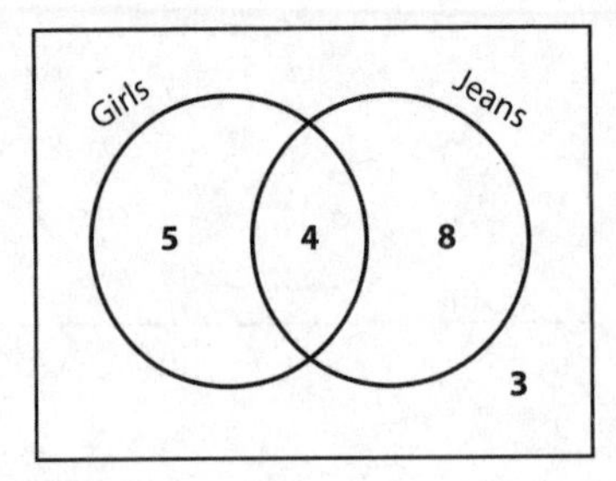

	Girls	Boys	Total
Jeans	4	8	12
Other	5	3	8
Total	9	11	20

Notice that the two-way table also has a row and a column for the individual totals as well as the grand total of items in the study. This becomes useful in looking at relative frequencies (that is, "what percent of the total falls under a particular category?").

Actually, you can find the percent of a single category compared to the total, as well as the percent of a specific combination as compared to the total. For example, $\frac{4}{9}$ (or about 44%) of females were wearing jeans, but $\frac{4}{20}$ (or 20%) of total students surveyed were jeans-wearing females.

From that small collection of data, many conclusions can be drawn: It is more common for a guy to wear jeans than a girl; there are more boys than girls in the class; more than half of the class was wearing jeans. These are just a few examples of what can be seen from this densely-packed array of information.

It is very easy to accidentally use the wrong denominator to answer a question from a two-way table. Be sure to pay close attention to the "out of" variable. One question may ask for a frequency to be compared to the row's or the column's total, while a different question asks for the frequency to be compared to the total number of things being analyzed.

Of course, all of these questions *could* have been answered using the Venn diagram and a little logic. But wasn't it so much easier with the information arranged in the two-way table?

Scatterplots: How Do You Determine the Line of Best Fit?

It is not often that something in life follows a perfect formula every single time. In the cases where it does happen to behave nicely, an equation can be used. The employer can figure out how much to pay her hourly staff, the grocer can calculate how much to charge for that bunch of bananas, and you can figure out how long it will take you to get to your destination at a law-abiding 55 miles per hour.

For the other 99% of the time, when various inconsistencies get in the way, hope is not lost. If the correlation between the two variables is strong enough, the data will still *resemble* a straight line when plotted on a plate. Plotting a series of ordered pairs from bivariate data results in a *scatterplot*.

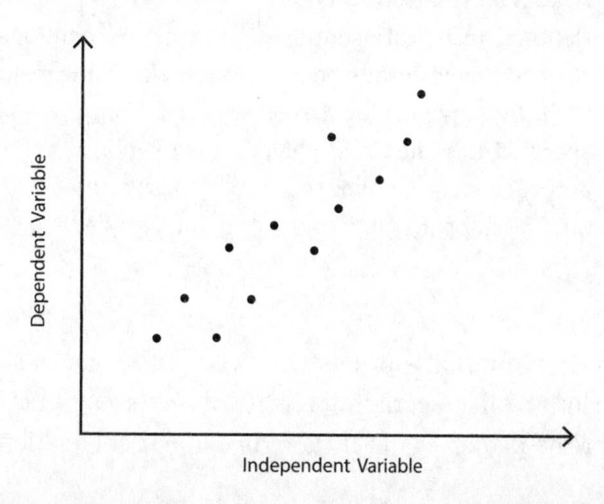

The placement of the variables on the two axes is very important. Considering the cause and effect of the two variables, the *effect* is the dependent variable, and is plotted on the vertical y-axis. The *cause* is the independent variable, and is plotted on the horizontal x-axis.

A scatterplot is a perfect name for these graphs. The points that you plot will appear, at first, scattered over the plane. However, if there is a correlation between the two variables, the scattered points will trend toward a line.

The stronger the correlation, the closer the points will be to forming a perfectly straight line. If the points trend toward a line but not tightly, it is said that there is a weak positive or weak negative correlation, depending on the direction of the line—upward for positive, downward for negative. In other words, the more scattered the points, the weaker the correlation.

The Line of Best Fit

Regardless if the correlation is strong or weak, a line is there . . . somewhere. Now, if a linear equation is given as a formula ahead of time, the points will obviously fall on a perfectly straight line. But you wouldn't call this a line of best fit. The graph wasn't actually a scatterplot; it was simply a graphed line. However, if a bivariate data set with any linear correlation at all is given, a line of best fit will exist.

Are all correlations linear?
No. To be technically sound, the correlations mentioned throughout this chapter *should* be labeled as linear. However, the majority of the work done in the eighth grade focuses on linear relationships and, unless otherwise specified, it is safe to assume that the relationships between the given variables are linear. Perhaps the second most common correlation is *exponential*. This happens when the output increases at an increasing rate.

Drawing a line of best fit on a scatterplot is somewhat of an art. It is essentially like averaging a picture. A series of points is literally scattered over the plane. Again, if there is a linear correlation between the variables, the points should gravitate toward a straight line . . . a straight line that isn't graphed yet.

The line of best fit should be drawn so that some points fall above it, some beneath it. Some points may end up directly on the line, but it is not necessary. The same number of points do not need to end up on one side of the line as the other. Wayward points (called outliers) may offset the numbers of points on each side of the line of best fit.

Consider the following attempts at drawing a line of best fit.

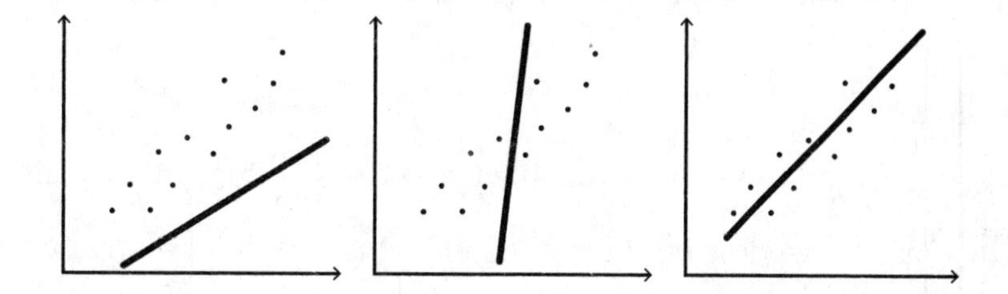

- The first line is drawn much too low. It does not accurately average out the points. It is not a good overall representation of the relationship between the two variables.
- The second line may divvy up the points well, but it is drawn at far too steep an angle to represent the true rate of change in the data set.
- The third example shows a well-drawn line of best fit. It averages the points well, and it is drawn with steepness consistent with the clustering of the points.

Finding the Equation of the Line of Best Fit

The graphed line, which represents the entire data set, can now be used a number of different ways. It could be used to compare rates of change with other scatterplots from similar data sets. It can also be used to make predictions about additional points from the same data set that have not yet been plotted.

To find the equation of the line, you need two things: the slope and the y-intercept. Since the drawn line was an estimate, it is also okay to estimate when finding these values. The y-intercept is the quick one. Estimate the value where the line of best fit crosses the y-axis. Record this value as b.

Finding the slope is a little bit more involved. Start by finding any two values that lie on the line of best fit. They may be close together, or far apart. The most important thing is that the two points be as close to the line of best fit as possible.

It is unnecessary for the two chosen points to be in the original data set. In fact, it is quite rare. Remember, you are finding the equation of the line of best fit, and that line will not include every point in the data set.

Find the slope of the line using the $\frac{\text{rise}}{\text{run}}$ formula for calculating the rate of change. It is not uncommon for this value to be a nasty decimal. Rounding to two or three digits is fine. The $y = mx + b$ form is generally the easiest form to use. Substitute your two values in for m and b, and you've got yourself the equation of the line of best fit.

Sometimes, the line of best fit will be drawn in such a fashion that you cannot tell what the y-intercept is. In these cases, you would have to use another form of linear equations called *point-slope form*: $y - y_1 = m(x - x_1)$. Here, m still means the slope but now, instead of being tied to using the y-intercept, you can use *any* point (x_1, y_1) on the line.

Once you have the equation of your line of best fit, you can plug in values of x to make predictions for an output. You can also plug in values of y to make predictions for what input would be necessary in order to obtain a desired output. And, of course, you can use the information within the equation to compare and contrast a scatterplot with other scatterplots, and make conclusions regarding the data with which you are working.

Practice Problems

Classify the following scatterplots into one of the following five categories:

- Strong positive correlation
- Strong negative correlation
- Weak positive correlation
- Weak negative correlation
- No correlation

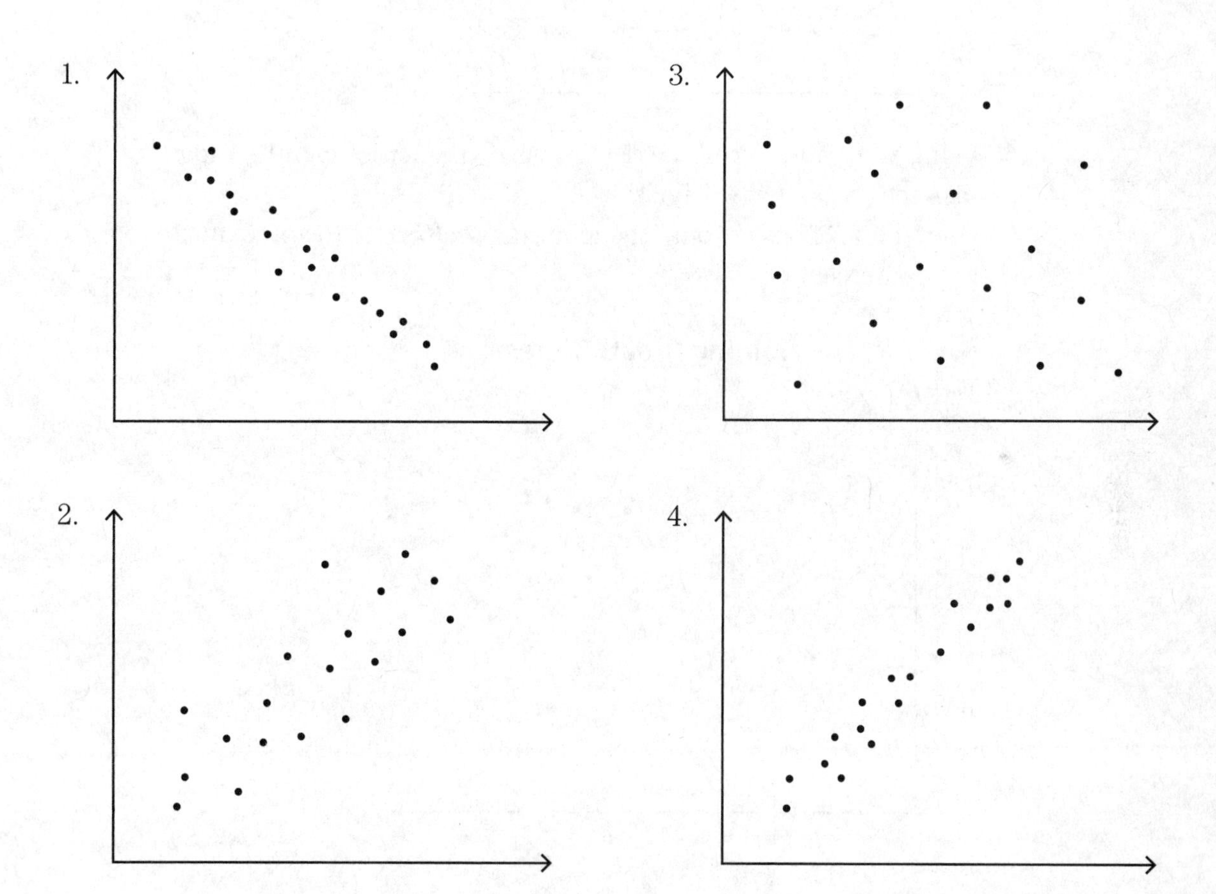

5. Use the given Venn diagram to create a two-way table.

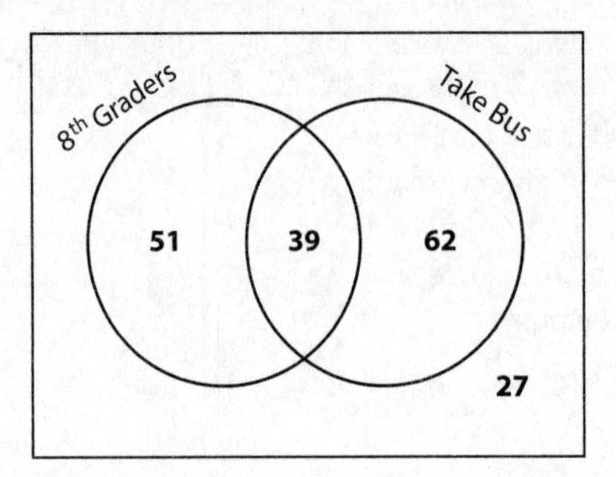

6. Using your work from question 5, what fraction of eighth graders take the bus to get to school?
7. Draw a line of best fit onto the following scatterplot to approximate the y-intercept.

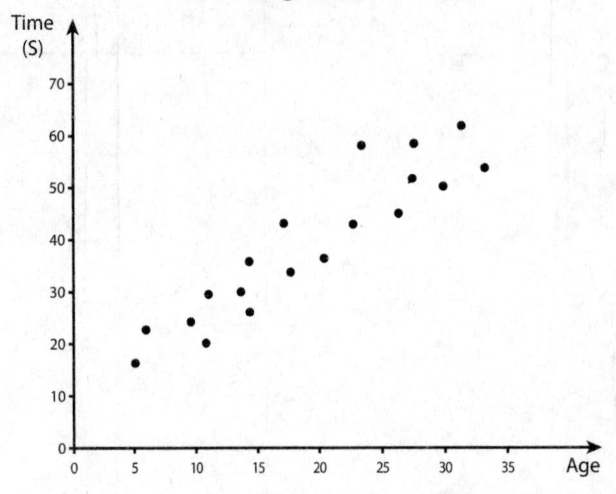

APPENDIX A

Answers to Practice Problems

Chapter 5

Write a ratio in simplest form to represent each scenario. Identify this as a part-to-part ratio or a part-to-whole ratio.

1. 5:3, $\frac{5}{3}$, or 5 to 3; part-to-part
2. 2 out of 5 students are boys; $\frac{2}{5}$, 2 to 5; part-to-whole
3. 3:4, $\frac{3}{4}$, 3 to 4; part-to-part
4. $\frac{6}{1}$, 6 to 1, 6:1; part-to-whole

Calculate the unit rate.

5. $1.12 per cupcake
6. $0.25 per ounce
7. 11 minutes per mile
8. $4.50 per pound
9. 472 miles per hour

Use ratio reasoning to solve word problems.

10.

Minutes	6	12	18	72	90	120
Miles	1	2	3	12	15	20

11. 24 apples will cost $6.00.
12. 337.5 people out of 1350 students will buy a hot lunch, but since you cannot have a half of a person, approximately 337 people will buy hot lunch.
13.

Noah							
Jada							

After Jada gives 21 sweatshirts to Noah, they have shared the work evenly.

Noah	7	7	7	7
Jada	7	7	7	7

Jada would have needed to sell 49 sweatshirts by herself before sharing the work with Noah.

14. Caleb did 21 crunches.

Calculate the percent of the quantity using any strategy.

15. 30
16. 9.2
17. 90
18. 2.5
19. 38 questions

Find the unknown value.

20. 75
21. 84
22. 10
23. 90

Convert between each of these units of measurement.

24. $8\frac{1}{2}$ gal = 34 qt
25. 25 ft = 300 in
26. 150 sec = 2.5 min

27. $2\frac{1}{4}$ tons = 4500 lbs

28. $3\frac{1}{3}$ yd = 10 ft

29. 12 lbs = 192 oz

Chapter 6

Determine the quotient of each of the problems.

1. $1\frac{1}{3}$
2. $10\frac{2}{3}$
3. $3\frac{3}{4}$
4. $2\frac{1}{4}$
5. 325
6. 15
7. 14
8. 354

Determine the sum or difference.

9. 81.34
10. 97.998
11. 45.16
12. 761.68

Determine the product.

13. 1.71
14. 2.475
15. $17.80

Determine the quotient.

16. 0.54
17. 34.5
18. 4.375

Determine the GCF or LCM for each question.

19. GCF(12, 32) = 4
20. LCM(2, 8) = 8
21. Three possible values for the missing number are 9, 18, and 36.
22. Anna could charge $9.00 for each bag.

Write an integer to represent each of the situations.

23. –30
24. 18
25. –25
26. 25

Determine the absolute value.

27. 15
28. 18
29. 6
30. $5 + 7 + 12 = 24$

Write an ordered pair to represent the location described.

31. (–3, –5)
32. (6, –4)
33. (1, 4)
34. (2, 4)
35. (–4, 5)
36. 16 units, (2, –8)

Chapter 7

Evaluate each of the following.

1. 25
2. 41
3. 24

Write each expression using exponents.

4. $2^2 \bullet 3^3 + 4^2 - 5$
5. $3^3 \bullet 1$ or $3^3 \bullet n^0$, where n is any number greater than or equal to 1.

Write each verbal phrase as an algebraic expression.

6. $6n + 5$
7. $n - 2$
8. $4(n - 3)$
9. $n/7 - 9$

Evaluate each of the following expressions.

10. 20
11. 24
12. 2
13. 98

Identify the property being used.

14. Commutative Property of Addition
15. Associative Property of Multiplication
16. Commutative Property of Multiplication
17. Identity Property of Multiplication also known as the Multiplication Property of One
18. Distributive Property
19. Commutative Property of Addition

Which value from the set is a solution to the given equation?

20. $x = 4$
21. $y = 10$
22. $p = 6$

Solve for the variable in each equation.

23. g = 16
24. $h = 40$
25. $k = 9$
26. $d = 72$

Write an inequality to represent each situation and then graph this on a number line.

27. i = thickness of ice; $i \geq 4$in
28. s = score to pass test; $s \geq 60\%$
29. p = number of people; $p \leq 15$
30. s = speed; $s \leq 65$ mph

Create a function table for each of the following situations. Identify the independent and dependent variable.

31.

Number of Songs	Total Cost
1	$12.00
2	$14.00
10	$30.00
30	$70.00

The expression to figure out the total cost is $2n + 10$, where n = the number of songs downloaded.

32.

Number of Hours	Distance Traveled in Miles
1	35
2	70
3	105
5	175

You will have traveled 175 miles.

Chapter 8

Calculate the area.

1. The area of the parallelogram is 14.7 in^2.
2. The area of the triangle is 30.875 m^2.
3. The area of the trapezoid is 47.5 ft^2.
4. The area of the composite figure is 89 yd^2.

Answer the following questions about surface area and nets.

5. The surface area is 72 ft^2.
6.

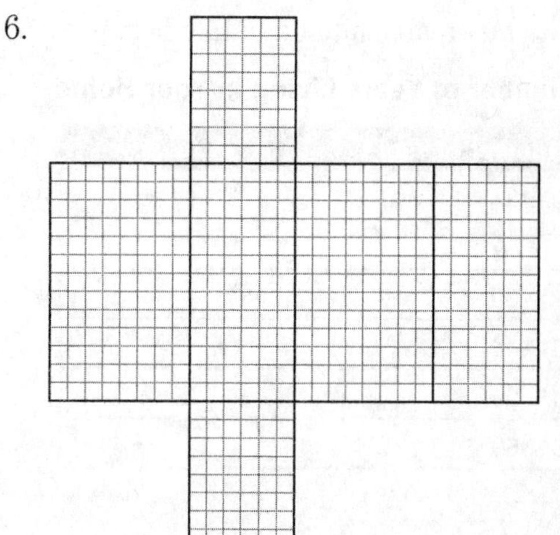

The surface area of this rectangular prism is 460 cm^2.

Calculate the volume.

7. It will take 2,898 $\frac{1}{4}$-unit cubes to make a rectangular prism that is $3\frac{1}{2}$ units by $5\frac{3}{4}$ units by $2\frac{1}{4}$ units in size.
8. The volume is 45.25 units^3.
9. 3.375 in^3.
10. The height of the storage unit is 10 ft.

Chapter 9

Using the sample data provided, answer each of the following questions.

A survey was completed that asked 25 random people how many years they have lived in their current home. Here is a list of the results.

Data: 5, 2, 7, 24, 53, 18, 13, 5, 2, 11, 14, 25, 36, 2, 6, 8, 15, 28, 30, 1, 15, 18, 21, 25, 16

1. The mean is 16.
2. The median is 15.
3. The mode is 2.
4. The range is 52.
5. The mean absolute deviation is 9.4.
6. **Number of Years Living in Your Home**

Number of Years	Frequency
0–10	9
11–20	8
21–30	6
31–40	1
41–50	0
51–60	1

Number of Years Living in Your Home

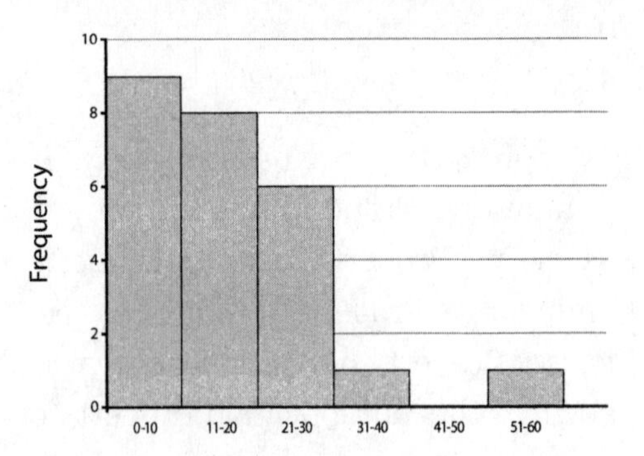

7. The lower quartile is 5.5.
8. The upper quartile is 24.5.
9. The interquartile range is 19.
10. 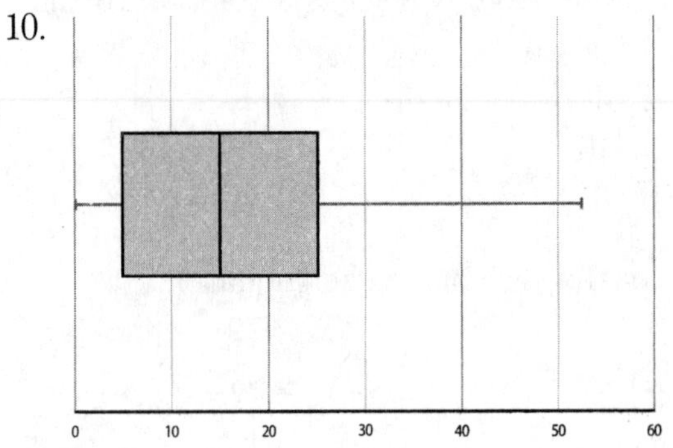

Chapter 11

1. $2\frac{4}{5}$ cups
2. 140 people
3. No
4. Yes
5. 32.5%
6. $390
7. 20% decrease
8. $30
9. 12.5% decrease

10. Tina had a better estimate. (Scott: 17.5% margin of error, and Tina: 15.625% margin of error)

Chapter 12

1. 5
2. 13
3. 5
4. −13
5. −11
6. 13
7. −36
8. 8
9. 52
10. $\frac{9}{5}$
11. $\frac{8}{11}$ is a repeating decimal. ($0.\overline{72}$)
12. $\frac{9}{80}$ is a terminating decimal. (0.1125)

Chapter 13

1. Expression
2. Equation
3. Expression
4. Equation
5. Equation
6. Expression
7. Equation
8. $5x$
9. $2x^2+5x$ (It's already completely simplified!)
10. $2x+8y$
11. $5x^2+3x+4$
12. $5x+20$
13. x^2-6x
14. $-3y+15$
15. $2z+8$
16. $5d-8$
17. 14 dogs
18. $3x+4 \leq 19$
19. $x \leq 5$
20. 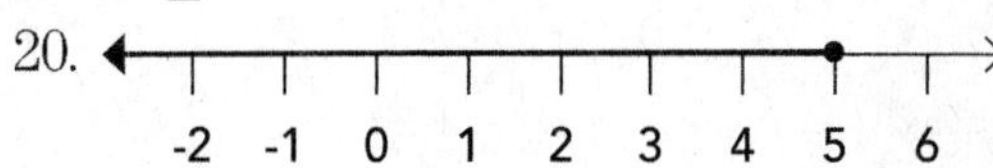

Chapter 14

1. Top = 150 feet, Bottom = 300 feet, Left = 200 feet, Right = 225 feet
2. 31.4 in
3. 200.96 cm^2
4. $x=85°$
5. 35°
6. $x=95°$
7. 96 ft^2

Chapter 15

1. 4
2. 1, 2, 3, 4, 5, 6, 7, 8, 9, 10
3. Very likely
4. $\frac{1}{2}$ or 0.5
5. $\left(\frac{4}{52}\right)\left(\frac{16}{51}\right)=\frac{64}{2652}$ or ≈ 0.0241
6. 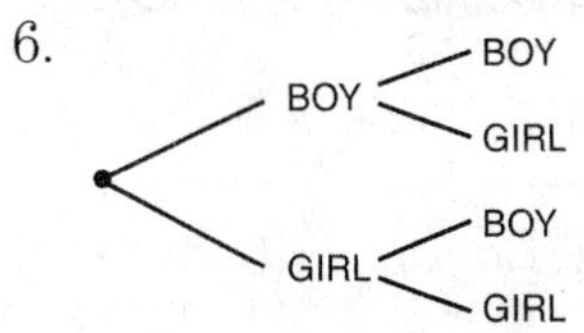

7. $8\times7\times6\times5\times4\times3\times2\times1=40{,}320$ ways

Chapter 17

1. -4, 3.6, 0, $\sqrt{81}$, $\frac{-12}{7}$
2. 6

3. 8
4. −3
5. 5 and 6
6. 11 and 12
7. 13 and 14
8. 8.72
9. 14.66
10. 22.36

Chapter 18

1. 1
2. $\frac{1}{25}$
3. 128
4. e^{10}
5. $3a^3b^8$
6. w^{15}
7. f^9
8. $x = \pm 12$
9. $x = \pm 20$
10. $x = 4$
11. $x = -1$
12. 51,300
13. 0.000000108
14. 4.6×10^9
15. 4.59×10^{-4}
16. $15 per hour
17. $m = 2$, $b = 2$
18. $m = -\frac{2}{3}$
19. $x = 3$
20. $(x, y) = (3, 55)$

Chapter 19

1. No
2. Yes
3. No
4. −45
5. 18 (Remember, exponents come before multiplication!)
6. 19
7. $x = -2$
8. 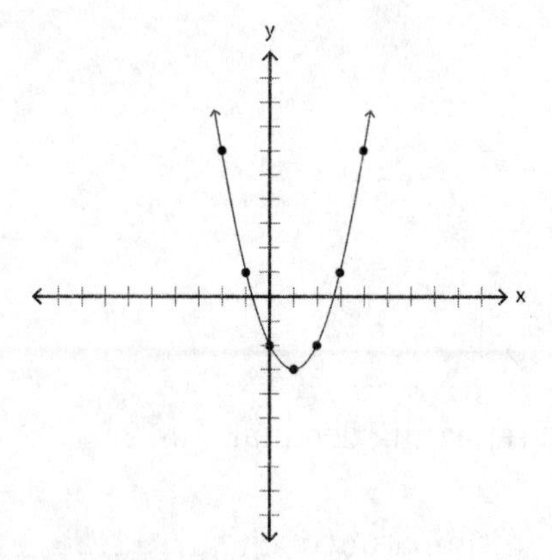

Chapter 20

1. 60 cm^2
2. 84 in^2
3. 80 mm^2
4. 452.16 ft^2
5. 153.86 cm^2
6. 900 cm^3
7. 480 in^3
8. 452.16 mm^3
9. 602.88 ft^3
10. 904.32 in^3
11. 256 in^3
12. 41.87 cm^3
13. 17 inches
14. 40 mm

15.

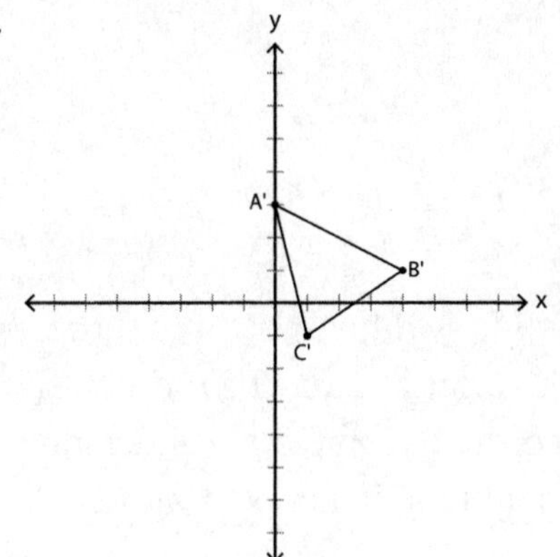

16.

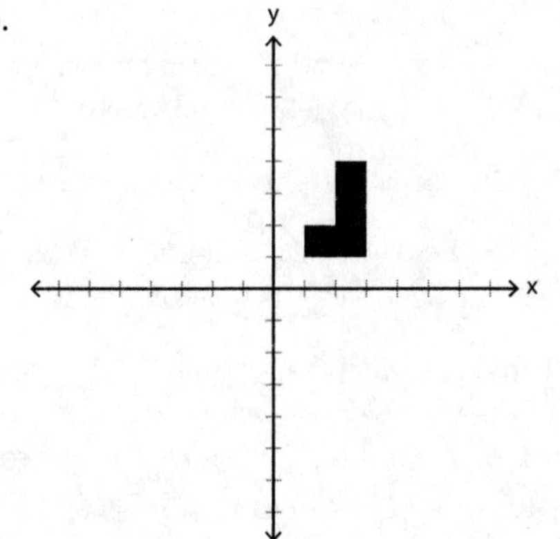

17.

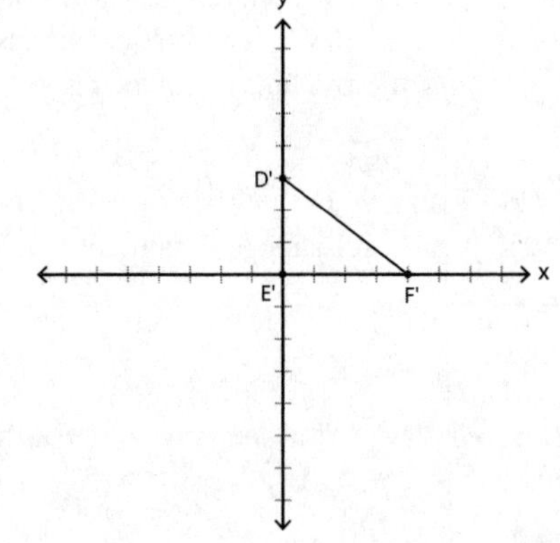

18.

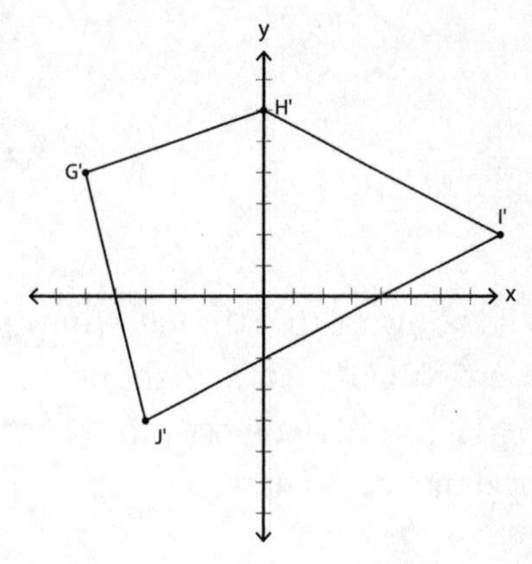

Chapter 21

1. Strong negative
2. Weak positive
3. No correlation
4. Strong positive
5.

	Bus	Other	Total
8th	39	51	90
Other	62	27	89
Total	101	78	179

6. $\frac{39}{90}$ (or 13 out of every 30) students.
7. **Holding Breath Times**

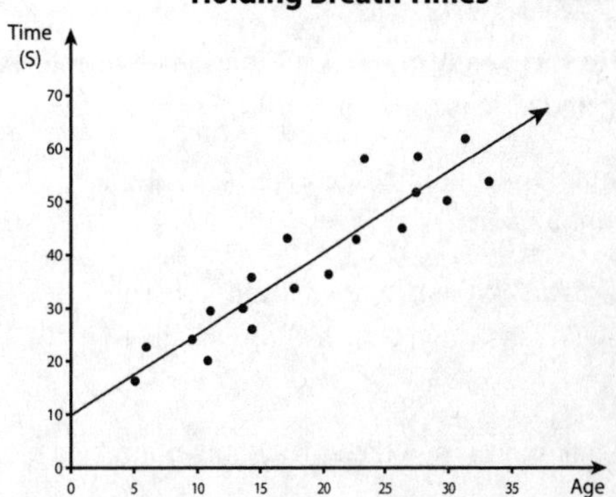

APPENDIX B

Glossary

You should be aware that the following definitions might appear slightly differently from those in whichever textbook your student uses at school. Although slight variations may be present, the ideas are the same. The most valuable thing a student can do in order to completely understand math vocabulary is to first learn the term and then rewrite the definition in her own words.

Absolute Value The distance between a given value and zero on the number line

Acute Angle Any angle with a measure between (but not including) 0 degrees and 90 degrees

Addend One of the values in an addition problem

Additive Identity Property of Zero The additive identity is the number zero. It is the number that has no effect when added to another number

Additive Inverses Two numbers whose sum is 0 are additive inverses of one another

Adjacent Angles Two angles which are side by side; they share a vertex and a ray, and do not overlap

Algebra The branch of mathematics in which symbols are used to represent numbers and express mathematical relationships

Algebraic Expression A combination of variables, numbers, and at least one operation

Algorithm A specific process or list of steps necessary to carry out a task

Altitude Also known as the height, it is the perpendicular distance from a shape's base to its furthest point

Area The amount of a flat surface taken up by a closed shape

Array A rectangular-shaped picture or arrangement of numbers often used to help model multiplication

Ascending An arrangement from least to greatest

Association Describes the cause-and-effect relationship between two variables (also see "Correlation")

Associative Property Allows a regrouping of either addends in an addition problem or factors in a multiplication problem, but does not change the end result; $a+(b+c)=(a+b)+c$, and $a(bc)=(ab)c$

Axis (Axes, plural) A straight line used as a reference for location; the vertical and horizontal lines that make up the quadrants of a Cartesian plane. The vertical axis is referred to as the y-axis and the horizontal axis is referred to as the x-axis

Axis of Reflection A straight line that serves as the line of symmetry between an image and its reflection

Base (as in 2D geometry) The length across the bottom of a shape

Base (as in 3D geometry) The shape on which a three-dimensional solid sits

Base (as a numerical value) The value(s) affected by an exponent; the number raised by an exponent

Benchmark Ratio A fraction (or decimal or percent) that is common enough that all three ratio forms should be memorized

Bivariate Data that is collected from two different variables; for example, a list of heights and weights of players on a football team

Box-and-Whisker Plot (also Box Plot) A visual representation in one-variable statistics, which summarizes the maximum, minimum, median, and quartiles of a data set, wherein the box shows the middle 50% of the data

Brackets Used when parentheses are around other parentheses; also known as braces

Cartesian Plane An infinitely large plane constructed by a horizontal x-axis and a vertical y-axis; also known as Coordinate Plane

Circle An infinite set of points all being the same distance away from a central point

Circumference The distance all the way around a circle

Closed Shape Any image on a flat surface with a specific interior region; all sides connect

Coefficient A number written in front of a variable part; it is "attached" by multiplication

Column A vertical arrangement of values in an array

Combination A collection of items where order does not matter

Combinatorics A branch of mathematics involving the selection and ordering of items

Common Core State Standards A set of national standards designed to guide instruction at each grade level, K–12

Commutative Property Allows the addends in an addition problem or the factors in a multiplication problem to be rearranged and the result to remain unchanged; $a+b=b+a$ and $ab=ba$

Complement (of an angle) The difference between a given angle and 90 degrees

Complementary Angles Angles whose measures add up to 90 degrees

Complex Fraction A fraction A/B, where A and/or B are fractions and B is not equal to zero

Composite Number A number that has more than two factors

Compound Probability A particular probability problem involving two, separate events

Cone A geometric solid consisting of one circular base and one vertex

Congruent Geometry's version of the word "equal"

Conjecture An educated guess based on mathematical reasoning

Consecutive One immediately following another

Constant A term with no variable part; a number

Conversion Factor A ratio written in fraction form that expresses the same value or quantity in two different units

Correlation Describes the cause-and-effect relationship between two variables (also see "Association")

Counting Number Positive numbers with no fractional or decimal parts; also called "natural numbers"

Cross Products The two products created from multiplying the values appearing diagonally in a proportion; if the two ratios are equal, the cross products will be equal

Cube A specific rectangular prism where the length, width, and height are all equal

Cube Root (of a number) The value that, when raised to the third power, results in the given number

Cylinder A geometric solid with two circular, parallel faces

Decimal One of three common forms of a ratio; a decimal is compared to one whole

Denominator The bottom number of a fraction; it represents the number of equal-sized sections in which the "whole" is divided

Dependent Events A situation where the result of the first event does affect the individual probabilities within the following events

Dependent Variable In bivariate data, this is used as the output value. It depends on the input as well as other varying factors

Descending An arrangement from greatest to least

Dilation A scaling up or down of an image; dilations preserve angle measures and change lengths proportionally

Diameter The total distance across a circle, through its center

Difference The result of a subtraction problem

Distribution The spread of a numerical data set over a dot plot, histogram, or other visual diagram

Distributive Property A multiplication of a value over at least two terms that have been added or subtracted; $a(b+c)=ab+ac$, $a(b-c)=ab-ac$

Dividend The number in a division problem appearing in the numerator of the fraction

Divisor The number in a division problem appearing in the denominator of the fraction

Domain The set of input values

Edge A line segment connecting two polygon faces of a geometric solid

Elimination One of the three most common methods used to solve a system of equations

Equation A complete number sentence where two expressions, separated by the symbol =, have the same value

Equilateral Triangle A triangle that has three sides of equal length and three congruent angles

Equivalent Expressions that, although they may not appear equal at first glance because they are written differently, have the same value. Example: $6(2x+5)=12x+30$

Event An action with a finite amount of known, possible conclusions

Experimental Probability The ratio of successful outcomes as compared to the number of trials conducted

Exponent A number attached to the top right of a value, denoting repeated multiplication

Expression A collection of numbers, variables, and/or operations

Face A flat surface of a geometric solid

Factor (as a noun) One of the values in a multiplication problem

Factor (of a given value) A number that divides nicely into the given value

Factor (as a verb) To split a value into at least two separate values that multiply together to the original value

Finite An amount of something that eventually ends

Fraction One of the three common forms of a ratio; it compares two numbers by separating them with a bar, taking the form a/b

Frequency How often something occurs

Function A set of ordered pairs where every single input value gets matched up with exactly one output value

Function Notation Where a set of operations is given and then potentially performed on input values

Function Rule The set of operations performed on the input values in order to get output values

Graph A visual representation of a solution or a set of solutions to an equation or inequality

Greatest Common Factor (GCF) The largest value that is a factor of every single one of the given values

Histogram A common graph used in statistics, similar to a bar graph, but instead of categories, the labels on the x-axis are numerical intervals

Horizontal The orientation of a line that points perfectly to the left and to the right

Hypotenuse The largest side of a right triangle; it is always located across from the right angle

Imaginary Number Result when the square root of a negative number is taken (although possibly mentioned in class, this is not in the middle-school Common Core)

Improper Fraction A ratio where the numerator is greater than the denominator

Independent Events A situation where the result of the first event does not affect the individual probabilities within the following events

Independent Variable In bivariate data, this is used as the input value; the "cause" in the "cause-and-effect" relationship

Inequality A comparison of two expressions that are not necessarily equal

Inference To make a conclusion from information presented

Infinite A never-ending amount of something

Interquartile Range In one-variable statistics, the difference between the first and third quartiles

Integer A number (positive, negative, or zero) with no fractional or decimal part

Inverse Operations Two operations that undo the effect of one another

Irrational A number that, as a decimal, does not terminate nor repeat

Isosceles Triangle A triangle with at least two sides of equal length

Kite A geometric shape with two distinct pairs of adjacent, congruent lengths

Least Common Denominator The smallest value that is a multiple of every single given denominator

Least Common Multiple (LCM) The smallest value that is a multiple of every single one of the given values

Leg Either of the two smaller sides of a right triangle; the two legs of the triangle form the right angle

Like Terms Any terms within an expression that have the same variable part

Line An infinite set of connected points

Linear A function that produces a straight line for a graph

Line Plot A method of visually displaying a distribution of data values where each data value is shown as a dot or mark above a number line; also known as a dot plot

Line Segment A finite set of connected points, possessing two endpoints; a subset of a line

Magnitude Another name for absolute value; it is the distance of a number from zero on the number line

Manipulatives Tangible items used to help model a mathematical problem

Margin of Error A way to express the amount of error in a guess or experiment $= \frac{\textit{Amount of Error}}{\textit{Actual Amount}} * 100\%$

Mean Also known as the average; the sum of all numbers divided by how many numbers there are

Mean Absolute Deviation The average of the differences between every individual piece of data and the mean of the data itself

Measures of Central Tendency Statistics used to describe a set of numerical data; they include but are not limited to the mean, the median, and the mode

Median The value that appears in the center of an ascending list of numbers; if there is an even amount of numbers, the average of the center two numbers is taken

Mixed Number A simplified form of an irrational number; it contains a whole amount as well as a fractional amount

Mode The number that appears most often in a set

Multiple (of a number) The product of a whole number and the given number

Multiplicative Identity The multiplicative identity is the number that has no effect when multiplied with another number; the number 1

Natural Number A positive number with no fractional or decimal part; also known as the "counting numbers"

Net A design created from an unfolded geometric solid

Number Line A single axis (usually horizontal) marked off with equally spaced numbers

Numerator The top number of a fraction; it refers to how many of the equal-sized pieces are being considered

Obtuse Angle Any angle with a measure between (but not including) 90 degrees and 180 degrees

Open Equation Any equation containing at least one variable

Opposite (of a number) The number needed to add to the given number in order to get a sum of zero; it essentially turns a positive to negative and a negative to positive

Order of Operations The order in which individual operations are performed when simplifying an expression (Parentheses, Exponents, Multiplication, and Division from left to right, Addition and Subtraction from left to right)

Ordered Pair A representation of the value of two variables written inside parentheses with the two values separated by a comma: (x, y)

Orientation The direction that an image is facing

Origin The point (0, 0) on the Cartesian plane (a.k.a. the Coordinate Plane)

Outcome The result of an event

Outlier A piece of data that is considerably different from the majority of the values

Parabola A U-shaped graph, created from a quadratic equation

Parallel Two distinct lines drawn at the same slope; they will never intersect

Parallelogram A quadrilateral with two pairs of parallel sides and opposite angles congruent

Percent One of three common forms of a ratio; a percent is always compared to 100

Percent of Change A measurement to show the significance of change $= \frac{\textit{Amount of change}}{\textit{Original amount}} * 100\%$

Perfect Square Number Any number that is the result of multiplying an integer by itself

Perimeter The total distance around the outside of a polygon

Permutation A selection of items where order is important

Perpendicular Two lines that intersect at a right angle

Pi The ratio between a circle's circumference and its diameter; it is approximately 3.14

Place Value The position of a digit within the entire number

Point-Slope Form One of three common linear forms of an equation; $y - y_1 = m(x - x_1)$, where m is the slope and (x_1, y_1) is any point on the line

Polygon A closed shape made up of straight-line segments

Power Another name for an exponent

Prime (in geometry) The new point or points obtained from performing a transformation on an image

Prime Number Any number with exactly two factors: 1 and itself

Prime Factorization An equivalent expression of an integer consisting only of prime factors

Prism A geometric solid consisting of two parallel polygon faces, connected by rectangular faces

Probability The likelihood of something occurring on a scale of $0 - 1; = \frac{\textit{number of possible desired outcomes}}{\textit{number of total possible outcomes}}$

Product The result of a multiplication problem

Proportion An equation of two (or more) equal ratios, usually fractions

Pyramid A geometric solid consisting of a polygon base and triangular faces, meeting at a top point

Pythagorean Theorem A mathematical law stating that the sum of the squares of the two shorter lengths in a right triangle is equal to the square of the longest side: $a^2 + b^2 = c^2$

Quadrant One of four partitions of the Cartesian plane created by the intersection of two axes; quadrants are numbered counterclockwise, starting with the top right

Quadratic Equation An equation where the largest power to which the input is raised is two, such as $y = ax^2 + bx + c$

Quadrilateral A closed geometric shape with four sides

Qualitative A description of something based on its qualities or characteristics

Quantitative A description of something based on its quantity or size

Quartiles The values which occur 25% of the way across an ascending list (1st quartile) as well as 75% of the way across (3rd quartile)

Quotient The result of a division problem

Radius The distance from a circle's center to the circle itself; it is equal to half of its diameter

Random Sample A strategic collection of data from a larger population, which contains no biases

Range (as in functions) The set of output values

Range (as in statistics) The difference between the maximum and minimum values in a set of data

Rate A ratio comparing two different units of measurement

Ratio A comparison of two numbers

Rational Number Any number that can be expressed as a ratio of integers a/b or $-a/b$; as a decimal, rational numbers either terminate or repeat

Ray A subset of a line with exactly one endpoint

Real Number Any number, either rational or irrational, but not imaginary

Reciprocal The number needed to multiply the given number by in order to get a product of 1; in fraction form, a reciprocal "flips" the fraction upside down

Rectangle A parallelogram with four right angles

Reflection A transformation that "flips" an image over a designated axis

Regular Polygon A polygon with all side lengths equal and all angles congruent

Relation A set of ordered pairs

Relative Frequency The ratio between the amount of a specific item's frequency and the total number of pieces of data

Repeating Decimal A rational number whose decimal part has a digit (or multiple digits) that recur(s) infinitely

Rhombus A quadrilateral with four equal sides and opposite angles congruent

Right Angle An angle with a measure of exactly 90 degrees

Right Triangle A triangle that contains a right angle

Rotation A transformation where an image is "spun" around a specific point of rotation by a given amount of degrees in a given direction

Row A horizontal arrangement of values in an array

Sample Space The list of all of the possible outcomes of an event

Scale Factor (with equivalent fractions) The value by which a fraction a/b is scaled to form its equivalent c/d

Scale Factor (with transformations) The value by which all lengths are multiplied by in a dilation

Scalene Triangle A triangle with three sides all of different lengths

Scatterplot A representation of bivariate data on a Cartesian plane; points are often "scattered" over the plane but will tend toward a pattern if a correlation is present

Scientific Notation A way to represent very large numbers or very small decimals using significant digits and place value

Similar Two images where every pair of corresponding angles is congruent and every pair of corresponding sides is in the same proportion

Simple Probability A particular probability problem where only one event takes place

Slope The amount of change in the output variable for each increase of one in the input variable; if the graph is available, $\frac{rise}{run}$; if two points on the line are given, $\frac{y_2 - y_1}{x_2 - x_1}$

Slope-Intercept Form One of three common forms of a linear equation; $y = mx + b$, where m is the slope and b is the y-intercept

Solid A three-dimensional geometrical object that takes up a specific amount of space

Solution The value(s) of the variable(s) in an open equation (or inequality) that make it a true statement

Solve To find the value(s) of the variable(s) in an open equation (or inequality) that make it a true statement

Sphere A geometric solid with no flat face or rigid edge; a ball

Square A quadrilateral with four equal sides and four 90-degree angles

Square Root (of a number) The value which, when multiplied by itself, results in the original given number

Standard Linear Form One of three common forms of a linear equation; $Ax + By = C$

Statistics A branch of mathematics that uses data to try to make predictions about the future by looking back on the past

Straight Angle An angle whose measure is exactly 180 degrees; a straight angle forms a straight line

Subscript A small number written to the bottom right of a variable, used to differentiate it from a similar variable; subscripts do not have a quantitative effect on the variable

Substitution One of three common methods of solving a system of equations

Sum The result of an addition problem

Supplement (of an angle) The difference between an angle measure and 180 degrees

Supplementary Angles Angles whose measures add up to 180 degrees

Surface Area The total area of every face and/or outside surface of a three-dimensional solid

System of Equations A collection of at least two equations where each instance of the same variable has the same value throughout

Table A way to represent data, often with categories written at the top of each column

Tape Diagram A drawing that looks like a segment of a tape, used to illustrate number relationships

Term Each part of an expression; can be a variable, a number, or the product/quotient of numbers and variables; terms are separated in an expression by addition and/or subtraction

Terminating Decimal A rational number whose decimal form ends in repeated zeroes

Transformation A movement of an image by either flipping, sliding, spinning, or a combination thereof; transformations preserve congruence

Translation A transformation where the image is shifted or slid in a specific direction comprised of a vertical and a horizontal component

Trapezoid A quadrilateral with exactly one pair of parallel sides

Tree Diagram A visual representation of compound probability that maps out the possible outcomes after each individual event

Trial The execution of an event

Triangle A geometric shape with three sides and three angles

Two-Way Table A visual representation of data that compares and contrasts two mutually inclusive variables

Unit Fraction Any fraction with a numerator of 1

Unit Rate Any rate with a denominator of 1

Variable A symbol, usually a letter, which represents an unknown or a varying value

Venn Diagram A visual representation of data that compares and contrasts two mutually inclusive variables

Vertex A corner point of an angle, shape, or a solid

Vertical The orientation of a line that points perfectly up and down

Vertical Angles Two angles that appear opposite each other in an intersection of straight lines; vertical angles are always congruent

Volume The amount of space taken up by a geometric solid

Whole Number A non-negative number with no fractional or decimal part

***X*-Axis** The horizontal number line in a Cartesian plane

***Y*-Axis** The vertical number line in a Cartesian plane

***Y*-Intercept** The value where a graph intersects the y-axis; the value of the output when $x = 0$

Zero Pair The combination of one positive and one negative unit; the net sum is zero, canceling out the units

Index